Robust Cooperative Control of Multi-Agent Systems

Robust Cooperative Control of Multi-Agent Systems

A Prediction and Observation Perspective

Chunyan Wang
Zongyu Zuo
Jianan Wang
Zhengtao Ding

CRC Press
Taylor & Francis Group
Boca Raton London New York

CRC Press is an imprint of the
Taylor & Francis Group, an **informa** business

First edition published 2021
by CRC Press
6000 Broken Sound Parkway NW, Suite 300, Boca Raton, FL 33487-2742

and by CRC Press
2 Park Square, Milton Park, Abingdon, Oxon, OX14 4RN

Library of Congress Cataloging-in-Publication Data

Visit the Taylor & Francis Web site at
http://www.taylorandfrancis.com

and the CRC Press Web site at
http://www.crcpress.com

ISBN: 978-0-367-75822-6 (hbk)
ISBN: 978-0-367-75823-3 (pbk)
ISBN: 978-1-003-16414-2 (ebk)

Typeset in CMR10
by KnowledgeWorks Global Ltd.

Contents

Author Bios

Chunyan Wang is currently an Associate Professor in the School of Aerospace Engineering at Beijing Institute of Technology (BIT), China. He received the B.Eng. degree in automatic control from Dezhou University, Shandong, China, in 2006, the M.S. degree in control theory and control engineering from Soochow University, Jiangsu, China, in 2009, the M.Sc. degree in electrical and electronic engineering from the University of Greenwich, London, U.K., in 2012, and the Ph.D. degree in control systems from the University of Manchester, Manchester, U.K., in 2016. He was a Research Associate with the School of Electrical and Electronic Engineering, University of Manchester, from November 2016 to March 2018. He was an academic visitor at the Department of Automatic Control and Systems Engineering, University of Sheffield, in December 2018. Dr. Wang serves as an Associate Editor of Nonlinear Control (specialty section of Frontiers in Control Engineering) and has over 50 journal and conference publications. His current research interests include cooperative control and guidance of multi-agent systems, UAVs, intelligent control and robotics.

Zongyu Zuo received the B.Eng. degree in automatic control from Central South University, Changsha, China, in 2005, and the Ph.D. degree in control theory and applications from Beihang University (formerly, Beijing University of Aeronautics and Astronautics), Beijing, China, in 2011. He was an Academic Visitor at the School of Electrical and Electronic Engineering, University of Manchester, Manchester, U.K., from September 2014 to September 2015, and held an inviting associate professorship with the Mechanical Engineering and Computer Science, UMR CNRS 8201, Université de Valenciennes et du Hainaut-Cambrésis, Valenciennes, France, in October 2015 and May 2017. He is currently a full Professor with the School of Automation Science and Electrical Engineering, Beihang University. His research interests are in the fields of nonlinear system control, control of UAVs and coordination of multi-agent systems. Dr. Zuo serves as an Associate Editor of the *Journal of the Franklin Institute*, *Journal of Vibration and Control*, *International Journal of Aeronautical and Space Sciences* and the *International Journal of Digital Signals and Smart Systems*.

Jianan Wang is currently an Associate Professor in the School of Aerospace Engineering at Beijing Institute of Technology, China. He received his B.S. and M.S. in Control Science and Engineering from the Beijing Jiaotong University and Beijing Institute of Technology, Beijing, China, in 2004 and 2007, respectively. He received his Ph.D. in Aerospace Engineering at Mississippi State University, Starkville, MS, USA, in 2011. His research interests include cooperative control of multiple dynamic systems, UAV formation control, obstacle/collision avoidance, trustworthy networked systems and estimation of sensor networks. He is a senior member of both IEEE and AIAA.

Zhengtao Ding is a Professor in the Department of Electrical and Electronic Engineering at University of Manchester, U.K. He received the B.Eng. degree from Tsinghua University, Beijing, China, and the M.Sc. degree in systems and control and the Ph.D. degree in control systems from the University of Manchester Institute of Science and Technology, Manchester, U.K. He was a Lecturer with Ngee Ann Polytechnic, Singapore, for ten years. In 2003, he joined the University of Manchester, Manchester, U.K., where he is currently the Professor of Control Systems with the Department of Electrical and Electronic Engineering. He has authored the book entitled *Nonlinear and Adaptive Control Systems* (IET, 2013) and a number of journal papers. His current research interests include nonlinear and adaptive control theory and their applications. Prof. Ding has served as an Associate Editor for the *IEEE TRANSACTIONS ON AUTOMATIC CONTROL*, the *IEEE CONTROL SYSTEMS LETTERS*, *Transactions of the Institute of Measurement and Control*, *Control Theory and Technology*, *Mathematical Problems in Engineering*, *Unmanned Systems* and the *International Journal of Automation and Computing*.

Preface

In the control community, "multi-agent systems" is a term used to describe a group of agents which are connected together to achieve specified control tasks over a communication network. Cooperative control of multi-agent systems has emerged as an attractive area of research in the last two decades and has been widely used in the real world, including wheeled robotic systems, satellites, autonomous underwater vehicles (AUVs), spacecraft, unmanned aerial vehicles (UAVs), sensor network, surveillance, smart grids, etc. In the formulation of cooperative control, there are two types of methods in the literature: centralized method and distributed method. Compared to centralized method which uses a core cell to control and connect a group of agents, the distributed method, which uses the local information to design distributed controllers, brings a number of benefits. For example, distributed multi-agent systems are more tolerant to bad environments, since failure of one agent does not seriously effect the performance of the whole system. Consensus control is a fundamental problem in distributed cooperative control, since many applications are based on the consensus algorithm design.

Robust cooperative control means that multi-agent systems are able to achieve specified control tasks while remaining robust to both parametric and nonparametric uncertainties. In this book, we present a concise introduction to the latest advances in robust cooperative control design for multi-agent systems. It covers a wide range of key issues in cooperative control, such as communication and input delays, parametric model uncertainties, external disturbances, etc. Different from the existing works, we present a systematic prediction and observation approach to design robust cooperative control laws. This is one unique feature of the book.

This book consists of ten chapters. In Chapter 1, we first review the multi-agent systems and some related work on robust cooperative control. Then, some mathematical background, including matrix theory, stability theory and basic algebraic graph theory are provided. Chapter 2 introduces the stability analysis for a class of nonlinear systems with input delay and provides some fundamental knowledge which will be used in the following chapters. In Chapter 3, robust consensus control for uncertian linear multi-agent systems with input delay is investigated. In Chapter 4, H_∞ consensus control of general linear multi-agent systems with input delay is studied. Chapter 5 introduces robust consensus control of nonlinar multi-agent systems with input delay. In Chapter 6, we investigate the consensus disturbance rejection problem for Lipschtz nonlinear multi-agent systems with disturbance observer-based control (DOBC). In Chapter 7, consensus disturbance rejection problem of nonlinear multi-agent systems is studied via distributed predictive observer design. The results in Chapter 6 are extended to formation control problem in Chapter 8 for leader-follower multi-agent systems. In Chapter 9, fixed-time formation control of multi-robot systems with uncertain communication delays is investigated. Chapter 10 investigates the cascade struture predictive observer design for consensus control with applications to UAVs formation flying.

The authors are indebted to our colleagues, collaborators and students for their help through collaboration on the topics of this monograph. In particular, the authors would like to thank Professor Zongli Lin of University of Virginia for his support and collaboration in some previous joint journal publications, which are related to the topic in this book. We would like to extend our thanks to Lian Sun and Jingying Chen at Taylor & Francis for their professionalism. We wish to thank our families for their support, patience and endless love.

Finally, we gratefully acknowledge the financial support from the National Natural Science Foundation of China under grants 61803032, 61673034, 62073019, 61873031 and the Beijing Institute of Technology Research Fund Program for Young Scholars. This work is also partially supported by the Key Laboratory of Dynamics and Control of Flight Vehicles (Beijing Institute of Technology), Ministry of Education.

This monograph was typeset by the authors using LaTeX. All simulations and numerical computations were carried out in MATLAB®.

Beijing, China	Chunyan Wang
Beijing, China	Zongyu Zuo
Beijing, China	Jianan Wang
Manchester, UK	Zhengtao Ding

Chapter 1

Introduction and Mathematical Background

In this chapter, we first review the cooperative control of multi-agent systems. Then, we briefly review how robust cooperative control is motivated and promoted in the control community. Following that, we give an overview of this monograph. Finally, some mathematical background, including matrix theory, stability theory and basic algebraic graph theory are provided.

1.1 Multi-Agent Coordination

The concept of multi-agent coordination is initially inspired by the observations and descriptions of collective behaviour in nature, such as fish schooling, bird flocking and insect swarming [108]. In 1987, three simple rules, separation (collision avoidance), alignment (velocity matching) and cohesion (flock centring), are proposed by Reynolds [129] to summarise the key characteristics of a group of biological agents. After that, a simple model is introduced by Vicsek [152] in 1995 to investigate the emergence of self-ordered motion in systems of particles with biologically motivated interaction. The flocking behaviours are later theoretically studied in [58, 109, 139, 148], to cite a few.

1.1.1 Control Architectures

For multi-agent systems, different kinds of control architectures have been illustrated in literature. Most of them could be cataloged as centralized and decentralized schemes. In centralized system, a core unit which connects all the robots is available of global team knowledge. By collecting and managing the information, the core unit can coordinate the motion of all the robots and guarantee the achievement of the mission. In such approaches, the central unit plays a fundamental role because it operates the whole system, i.e., it has to receive the information obtained by all the sensors and make all the calculations, decisions and communications with all the robots. Thus, advanced and expensive equipments are necessary to satisfy all the technological requirements. For decentralized schemes, all the robots are in the same level and have the same equipments. Each robot uses the local sensor to obtain the relative state information of its neighbours, and then makes decision for the next step. Furthermore, each vehicle does not need the global information/mission and just communicates/shares information with their neighbour robots.

1.1.2 Potential Applications

Cooperative control has broad potential applications in real world including wheeled robotics system [31], satellites formation [124, 131], autonomous underwater vehicles (AUVs) [6, 112], spacecraft [102], unmanned aerial vehicles(UAVs) [1, 138], automated highway systems [81], sensor network [134], surveillance [127], smart grid [106] and so on. Typical tasks include consensus [110, 125], flocking [97], swarming [133], formation control [118], synchronization [69, 134], cooperative guidance [17, 154] and distributed filtering [15, 137]. Additionally, cooperative control of multiple mobile robots can implement some specified tasks such as distributed manipulation [98], mapping of unknown or environments [41,135], rural search and rescue [103], transportation of large objects [183, 186], etc.

1.1.3 Research Topics

- **Consensus Control**. Consensus control is a fundamental problem in cooperative control of multi-agent systems, since many applications are based on the consensus algorithm design. In multi-agent systems, consensus problem means how to design the control strategy for a group of agents to reach a consensus (or agreement) of the states of interest. The basic idea is that each agent updates its information state based on the information states of its local neighbours in such a way that the final information state of each agent converges to a common value [127]. One significant contribution in consensus control is the introduction of graph theory to the conventional control theory [21, 38, 102]. A general framework of the consensus problem for networks of single integrators is proposed in [110]. Since then, consensus problems have been intensively studied in many directions.

  In terms of the dynamics of agents, consensus problems for various systems' dynamics have been massively investigated. The system dynamics has large influence on the final consensus state of the multi-agent systems. For example, the consensus state of multi-agent systems with single integrator dynamics often converges to a constant value, and meanwhile, consensus for second-order dynamics might converge to a dynamic final value (i.e., a time function) [13]. Many early results on consensus problems are based on simple agent dynamics such as first or second-order integrator dynamics [12, 39, 53, 59, 125, 126]. However, in reality a large class of practical physical systems cannot be feedback linearised as first or second-order dynamical model. For instance, for a group of

UAVs [28], higher-order dynamic models may be needed. More complicated agent dynamics are described by high-order linear multi-agent systems in [55, 73, 74, 105, 142, 195]. After that, the results are extended to nonlinear multi-agent systems [26, 27, 76, 140, 175, 190]. Consensus for nonlinear systems is more involved than that for their linear systems counterparts. The difficulty of consensus control for nonlinear systems owes to certain restrictions the nonlinearity imposes on using the information of the individual systems. Consensus control for second-order Lipschitz nonlinear multi-agent systems is addressed in [191]. The consensus problems of high-order multi-agent systems with nonlinear dynamics are studied in [26, 76, 175, 190]. The works [27, 140] address the consensus output regulation problem of nonlinear multi-agent systems. A common assumption in the previous results is that the dynamics of the agents are identical and precisely known, which might not be practical in many circumstances. Due to the existence of the non-identical uncertainties, the consensus control of heterogeneous multi-agent systems is studied in [24, 61, 150, 196].

The communication connections between the agents are also playing an important role in consensus problems. Most of the existing results are based on fixed communication topology, which indicates that the Laplacian matrix $\mathcal{L}$ is a constant matrix (see Chapter 2 for graph theory notations). It is pointed out in [88, 122] that the consensus is reachable if and only if zero is a simple eigenvalue of $\mathcal{L}$. If zero is not a simple eigenvalue of $\mathcal{L}$, the agents cannot reach consensus asymptotically as there exist at least two separate subgroups or at least two agents in the group who do not receive any information [155]. It is also known that zero is a simple eigenvalue of $\mathcal{L}$ if and only if the directed communication topology has a directed spanning tree or the undirected communication topology is connected [125]. The results with directed graphs are more involved than those with undirected graphs. The main problem is that the Laplacian matrix associated with a directed graph is generally not positive semi-definite [71]. The decomposition method for the undirected systems cannot be applied to the directed one due to this unfavourable feature. For the consensus control with directed communication graphs, the balanced and/or strong connected conditions are needed in [75, 167], which are stronger than the directed spanning tree condition. In practice, the communication between the agents may not be fixed due to technological limitations of sensors or link failures. The consensus control of multi-agent systems with switching topologies has been investigated in [175, 177]. In [191] and [176], consensus control with communication constraint and Markovian communication failure are studied. In [48, 49, 144], event-triggered security output feedback control is applied for networked interconnected systems subject to cyber-attacks.

In term of the number of leaders, the above researches can also be roughly specified into three catalogs, that is, leaderless consensus (consensus without a leader) [26,73] whose agreement value depends on the initial states of the agents, leader-follower consensus (or consensus tracking) [53,175] which has a leader agent to determine the final consensus value, and containment control [12,59] where has more than one leader in agent networks. Compared to leaderless consensus, consensus tracking, and containment control have the advantages to determine the final consensus value in advance [142].

- **Formation Control**. Apart from consensus control which is to drive all agents reach the same desired value, another vital research direction is formation control, in which the agents form a predesigned geometrical configuration through local interactions. Compared to consensus control, the final states of all agents are more diverse under the formation control scenarios.

  Formation control is one of the most important applications in cooperative control. Different control strategies have been used for formation control, such as actual leader [18], virtual leader [35], behavioural [7], etc. A survey on various classifications of formation control is given in [107]. Consensus-based formation control strategies have also been investigated [32,122,123,128,141,194]. The role of Laplacian eigenvalues in determining formation stability is investigated in [39]. In [123], it is pointed out that consensus-based formation control strategies are more general and contain many existing actual/virtual leader and behavioural approaches as special cases. Most of the early results on formation control focus on simple agent dynamics such as first or second-order integrator dynamics. In reality, some practical physical systems cannot be feedback linearised as first or second-order dynamical model. Formation control for a class of high-order linear multi-agent systems with time delays are studied in [32]. The applications of formation control in various areas could be found in [122,141,194].

  By different types of sensing and controlled variables, the formation control problem can be classified into position-, displacement-, and distance-based control [107]. For agents that receive specified positions with respect to the global coordinate system, they only sense their own positions, and this is called position-based control. For agents that receive orientation of the global coordinate system, they sense relative positions of its neighbouring agents, and this is called displacement-based control. For agents that receive desired inter-agent distances, they sense relative positions of their neighbouring agents with respect to their own local coordinate systems, and this is called distance-based control.

- **Other Directions**. Except for consensus control and formation control, there are also some other directions that have been well studied, such as flocking control [109, 111, 129, 146, 147], rendezvous [2, 20, 82, 83, 151], synchronization [36,62,93,94,115–117,119,173,180], containment control [10, 11, 14, 22, 23, 40, 59, 77, 99, 100] and so on.

1.2 Robust Problem in Cooperative Control

Robustness of control systems to disturbances and uncertainties has always been a popular topic in feedback control systems. Feedback would not be necessary for most control systems if there are no disturbances and uncertainties [202]. In consensus control of multi-agent systems, modelling errors, external disturbances, time delays caused by communications and some other issues will significantly affect the consensus performance. Thus, robust consensus control problem has been developed as one of the most important topics in this area.

1.2.1 Time-Delay Problem

Systems with delays frequently appear in engineering systems. Typical examples of time-delay systems are communication networks, industrial processes, underwater vehicles and so on [197]. For multi-agent systems, time delays arising from agents are diverse. Communication delay is one source of delay due to the interactions between the agents. Another source of delay is in the input and output channel due to the decision-making and signal processing. In applications, for vision-based autonomous robots, camera latency and image processing will cause delays as well. The presence of delays, especially, long delays, makes system analysis and control design much more complicated. Therefore, the stabilization of time-delay systems has attracted much attention in both academic and industrial communities; see the surveys [47, 130], the monographs [46, 64, 197], and the references therein.

In the formulation of stabilization of time-delay systems, there are two types of feedback methods in the literature: standard (memoryless) feedback and predictive (memory) feedback, respectively. Memoryless controllers are useful for the systems with state delays [19, 52, 80, 182, 193]. However, it is known that system with input delay is more difficult to handle in control theory [130]. For predictive feedback, compensation is added in the controller design to offset the adverse effect of the time delay and the stabilization problems are reduced to similar problems for ordinary differential equations. A wide variety of predictor-based methods such as Smith predictor [136], modified Smith predictor [113], finite spectrum assignment [96] and Artstein-Kwon-Pearson reduction method [3, 66] are effective and efficient when the delay is too large to be neglected and a standard (memoryless) feedback would fail. However a drawback of the predictor-based methods is that the controllers involve integral terms of the control input, resulting in difficulty for the control implementation [37]. A halfway solution between these two methods is to ignore the troublesome integral part, and use the prediction based on the exponential of the system matrix, which is known as the truncated prediction feedback (TPF) approach. This idea started from low-gain control of the systems with input saturation [86], then it is developed for linear systems [87,189,200], and nonlinear systems [30,206].

Many existing works on consensus control have considered delays, [56,162, 166,171,192,198], to cite a few. Output feedback based algorithms are designed in [56] for multi-agent systems with state time delay. A PD-like protocol is designed in [166] to average consensus control for systems with time delay and switching topology. The existing studies [101,145,185,198] of consensus control with communication delays mainly focus on linear multi-agent systems. Consensus problems for nonlinear multi-agent systems with time delay are expected to be more complicated just as stabilization of a nonlinear system with time delay is much more involved than its linear counterpart. Additional care is required to tackle the influence of the nonlinearity that appears in the agent dynamics. In [161] and [162], predictor based and truncated prediction based method are proposed to deal with the input delay for Lipschitz nonlinear multi-agent systems, respectively.

1.2.2 Model Uncertanties and External Disturbances

The practical physical systems often suffer from uncertainties which may be caused by mutations in system parameters, modelling errors or some ignored factors [114]. The robust consensus problem for multi-agent systems with continuous-time and discrete-time dynamics are investigated in [51] and [50], where the weighted adjacency matrix is a polynomial function of uncertain parameters. Most of the existing results on consensus control of uncertain multi-agent systems are often restricted to certain conditions, like single or double integrators [84,149], or undirected network connections [91].

Other than the parameter uncertainties, the agents may also be subject to unknown external disturbances, which might degrade the system performance and even cause the network system to diverge or oscillate. The robust H_∞ consensus problems are investigated for multi-agent systems with first and second-order integrators dynamics in [84,85]. The H_∞ consensus problems for general linear dynamics with undirected graphs are studied in [72,92]. The results obtained in [72] are extended to directed graph in [167]. The H_∞ consensus problems for switching directed topologies are investigated in [132,178]. The nonlinear H_∞ consensus problem is studied in [75] with directed graph. Global H_∞ pinning synchronization problem for a class of directed networks with aperiodic sampled-data communications is addressed in [179]. It is worth noting that the directed graphs in [75,167] are restricted to be balanced or strongly connected. The main problem is that the Laplacian matrix associated with a directed graph is generally not positive semi-definite [71]. The decomposition method developed in [72] for the undirected systems cannot be applied to the directed one due to this unfavourable feature.

Sometimes, disturbances in real engineering problems are periodic and have inherent characteristics such as harmonics and unknown constant load [16]. For those kinds of disturbances, it is desirable by utilizing the disturbance information in the design of control input to cancel the disturbances directly. One common design method is to estimate the disturbance by using the measurements of states or outputs and then use the disturbance estimate to compensate the influence of the disturbance on the system, which is referred to as Disturbance Observer-Based Control (DOBC) [70]. Using DOBC method, consensus of second-order multi-agent dynamical systems with exogenous disturbances is studied in [34,184] for matched disturbances and in [171] for unmatched disturbances. Disturbance observer based tracking controllers for high-order integrator-type and general multi-agent systems are proposed in [5,10], respectively. A systematic study on consensus disturbance rejection via disturbance observers could be found in [28].

1.3 Overview of This Monograph

This book presents a concise introduction to the latest advances in robust cooperative control design for multi-agent systems with input delay and external disturbances, especially from a prediction and observation perspective. It covers a wide range of applications, such as trajectory tracking of quadrotors, formation flying of multiple UAVs and fixed-time formation of ground vehicles, etc.

The book is organized as follows. In Chapter 1, we first review the multi-agent systems and some related work on robust cooperative control. Then, some mathematical backgrounds, including matrix theory, stability theory and basic algebraic graph theory are provided. Chapter 2 introduces the stability analysis for a class of nonlinear systems with input delay and provides some fundamental knowledge which will be used in the following chapters. In Chapter 3, robust consensus control for uncertian linear multi-agent systems with input delay is investigated. In Chapter 4, H_∞ consensus control of general linear multi-agent systems with input delay is studied. Chapter 5 introduces robust consensus control of nonlinar multi-agent systems with input delay. In Chapter 6, we investigate the consensus disturbance rejection problem for Lipschtz nonlinear multi-agent systems with disturbance observer-based control. In Chapter 7, consensus disturbance rejection problem of nonlinear multi-agent systems is studied via distributed predictive observer design. The results in Chapter 6 are extended to formation control problem in Chapter 8 for leader-follower multi-agent systems. In Chapter 9, fixed-time formation control of multi-robot systems with uncertain communication delays is investigated. Chapter 10 investigates the cascade struture predictive observer design for consensus control with aplications to UAVs formation flying.

1.4 Mathematical Background

In this section, some mathematical notations and basic definitions that will be used in the remainder of this book are provided.

1.4.1 Notations

Throughout the book, $\mathbb{R}^{n\times m}$ and $\mathbb{R}^n$ represent a set of $n \times m$ real matrices and n-dimensional column vectors, and $0_{n\times m}$ denotes the $n \times m$ matrices with all zeros. Let $\mathbf{1}$ and I represent a column vector with all entries equal to one and the identity matrix with appropriate dimension, respectively. $\mathcal{L}_2^p\,[0,\infty)$ denotes the space of p-dimensional square integrable functions over $[0,\infty)$.Given a real matrix $X \in R^{n\times m}$ and vector $x \in R^n$, $\|X\|_F$ denotes the Frobenius norm, and $\|x\|$ is the Euclidean norm. The signal $\otimes$ denotes the Kronecker product of matrices, and the notation diag(π_i) denotes a block-diagonal matrix with $\pi_i, i = 1, \cdots, N$, on the diagonal. The matrix inequality $A > B$ means that $A - B$ is positive definite. $\otimes$ denotes the Kronecker product of matrices.

1.4.2 Matrix Theory

Definition 1 *The Kronecker product of matrix $A \in \mathbb{R}^{m\times n}$ and $B \in \mathbb{R}^{p\times q}$ is defined as*

$$A \otimes B = \begin{bmatrix} a_{11}B & \cdots & a_{1n}B \\ \vdots & \ddots & \vdots \\ a_{m1}B & \cdots & a_{mn}B \end{bmatrix},$$

and they have following properties:

1. $(A \otimes B)(C \otimes D) = (AC) \otimes (BD)$,
2. $(A \otimes B) + (A \otimes C) = A \otimes (B + C)$,
3. $(A \otimes B)^{-1} = A^{-1} \otimes B^{-1}$,
4. $(A + B) \otimes C = (A \otimes C) + (B \otimes C)$,
5. $(A \otimes B)^T = A^T \otimes B^T$,

6. *If $A \in \mathbb{R}^{m\times n}$ and $B \in \mathbb{R}^{p\times q}$ are both positive definite (positive semi-definite), so is $A \otimes B$,*

where the matrices are assumed to be compatible for multiplication.

Lemma 1.4.1 (Gershgorin's Disc Theorem [54]) *Let $A = [a_{ij}] \in \mathbb{R}^{n\times n}$, let*

$$R_i^{'}(A) \equiv \sum_{j=1, j\neq i}^{n} |a_{ij}|, i = 1, 2, \cdots, n$$

denote the deleted absolute row sums of A, and consider the n Gersgorin discs

$$\left\{z \in \mathbb{C} : |z - a_{ii}| \leq R_i^{'}(A)\right\}, i = 1, 2, \cdots, n.$$

The eigenvalues of A in the union of Gersgorin discs are given by

$$G(A) = \bigcup_{i=1}^{n} \left\{z \in \mathbb{C} : |z - a_{ii}| \leq R_i^{'}(A)\right\}.$$

Furthermore, if the union of k of the n discs that comprise $G(A)$ forms a set $G_k(A)$ that is disjoint from the remaining $n - k$ discs, then $G_k(A)$ contains exactly k eigenvalues of A, counted according to their algebraic multiplicities.

Definition 2 *[54] A matrix $A = [a_{ij}] \in \mathbb{R}^{n\times n}$ is diagonally dominant if*

$$|a_{ii}| \geq \sum_{j=1, j\neq i}^{n} |a_{ij}| = R_i^{'}(A), \ \forall i = 1, 2, \cdots, n.$$

It is strictly diagonally dominant if

$$|a_{ii}| > \sum_{j=1, j\neq i}^{n} |a_{ij}| = R_i^{'}(A), \ \forall i = 1, 2, \cdots, n.$$

Definition 3 (M-matrix, Definition 6 in [71]) *A square matrix $A \in \mathbb{R}^{n\times n}$ is called a singular (nonsingular) M-matrix, if all its off-diagonal elements are non-positive and all eigenvalues of A have nonnegative (positive) real parts.*

Lemma 1.4.2 (Schur Complement Lemma, [46]) *For any constant symmetric matrix*

$$S = \begin{bmatrix} S_{11} & S_{12} \\ S_{12} & S_{22} \end{bmatrix},$$

the following statements are equivalent:
(1) $S < 0$,
(2) $S_{11} < 0, \ S_{22} - S_{12}^T S_{11}^{-1} S_{12} < 0$,
(3) $S_{22} < 0, S_{11} - S_{12} S_{22}^{-1} S_{12}^T < 0$.

Lemma 1.4.3 (Finsler's Lemma, [71]) *For $P = P^T \in \mathbb{R}^{n\times n}$ and $H \in \mathbb{R}^{m\times n}$, the following two statements are equivalent:*
(1) $P - \sigma H^T H < 0$ holds for some scalar $\sigma \in \mathbb{R}$,
(2) There exists a $X \in \mathbb{R}^{n\times m}$ such that $P + XH + H^T X^T < 0$.

Lemma 1.4.4 ([46]) *For any given $a, b \in \mathbb{R}^n$, we have*

$$2a^T SQb \leq a^T SPS^T a + b^T Q^T P^{-1} Qb,$$

where $P > 0$, S and Q have appropriate dimensions.

Lemma 1.4.5 (Young's Inequality, [71]) *For nonnegative real numbers a, b, if p, q are real numbers that satisfy $\frac{1}{p} + \frac{1}{q} = 1$, then $ab \leq \frac{a^p}{p} + \frac{b^q}{q}$.*

1.4.3 Stability Theory

In this subsection, some basic concepts of stability based on Lyapunov functions are provided. The material in this subsection is mainly from [25].

Consider a nonlinear system

$$\dot{x} = f(x), \tag{1.1}$$

where $x \in \mathcal{D} \subset \mathbb{R}^n$ is the state of the system, and $f : \mathcal{D} \subset \mathbb{R}^n \longrightarrow \mathbb{R}^n$ is a continuous function, with $x = 0$ as an equilibrium point, that is $f(0) = 0$, and with $x = 0$ as an interior point of $\mathcal{D}$. $\mathcal{D}$ denotes a domain around the equilibrium $x = 0$.

Definition 4 (Lyapunov stability) *For the system (1.1), the equilibrium point $x = 0$ is said to be Lyapunov stable if for any given positive real number R, there exists a positive real number r to ensure that$\|x(t)\| < R$ for all $t > 0$ if $\|x(0)\| < r$. Otherwise, the equilibrium point is unstable.*

Definition 5 (Asymptotic stability) *For the system (1.1), the equilibrium point $x = 0$ is asymptotically stable if it is stable (Lyapunov) and furthermore* $\lim_{t\to\infty} x(t) = 0$.

Definition 6 (Exponential stability) *For the system (1.1), the equilibrium point $x = 0$ is exponential stable if there exist two positive real numbers α and λ such that the following inequality holds:*

$$\|x(t)\| < \alpha \|x(0)\| \exp^{-\lambda t},$$

for $t > 0$ in some neighbourhood $\mathcal{D} \subset \mathbb{R}^n$ containing the equilibrium point.

Definition 7 (Globally asymptotic stability) *If the asymptotic stability defined in Definition 5 holds for any initial state in $\mathbb{R}^n$, the equilibrium point is said to be globally asymptotically stable.*

Definition 8 (Globally exponential stability) *If the exponential stability defined in Definition 6 holds for any initial state in $\mathbb{R}^n$, the equilibrium point is said to be globally exponentially stable.*

Definition 9 (Positive definite function) *A function $V(x) \in \mathcal{D} \subset \mathbb{R}^n$ is said to be locally positive definite if $V(x) > 0$ for $x \in \mathcal{D}$ except at $x = 0$ where $V(x) = 0$. If $\mathcal{D} = \mathbb{R}^n$, i.e., the above property holds for the entire state space, $V(x)$ is said to be globally positive definite.*

Definition 10 (Lyapunov function) *If in $\mathcal{D} \in \mathbb{R}^n$ containing the equilibrium point $x = 0$, the function $V(x)$ is positive definite and has continuous partial derivatives, and if its time derivative along any state trajectory of system (1.1) is non-positive, i.e.,*

$$\dot{V}(x) \leq 0,$$

then $V(x)$ is a Lyapunov function.

Definition 11 (Radially unbounded function) *A positive definite function $V(x) : \mathbb{R}^n \longrightarrow \mathbb{R}$ is said to be radially unbounded if $V(x) \longrightarrow \infty$ as $\|x\| \longrightarrow \infty$.*

Theorem 1 (Lyapunov theorem for global stability, Theorem 4.3 in [25]) *For the system (1.1) with $\mathcal{D} \in \mathbb{R}^n$, if there exists a function $V(x) : \mathbb{R}^n \longrightarrow \mathbb{R}$ with continuous first order derivatives such that*

- *$V(x)$ is positive definite*
- *$\dot{V}(x)$ is negative definite*
- *$V(x)$ is radially unbounded*

then the equilibrium point $x = 0$ is globally asymptotically stable.

1.4.4 Basic Algebraic Graph Theory

In this subsection, some basic concepts of graph theory are provided. The materials in this subsection are mainly from some monographs, such as [68, 71, 122, 127, 168].

Basic definition

The graph theory has been introduced by Leonard Euler in year 1736. Generally it is convenient to model the information exchanges among agents by directed or undirected graphs. A directed graph $\mathcal{G} \triangleq (\mathcal{V}, \mathcal{E})$, in which $\mathcal{V} \triangleq \{v_1, v_2, \cdots, v_N\}$ is the set of nodes, and $\mathcal{E} \subseteq \mathcal{V} \times \mathcal{V}$ is the set of edges with the ordered pair of nodes. A vertex represents an agent, and each edge represents a connection. A weighted graph associates a weight with every edge in the graph. Self loops in the form of (v_i, v_i) are excluded unless otherwise indicated. The edge (v_i, v_j) in the edge set $\mathcal{E}$ denotes that agent v_j can obtain information from agent v_i, but not necessarily vice versa (see Fig. 1.1 (a)). A graph with the property that $(v_i, v_j) \in \mathcal{E}$ implies $(v_j, v_i) \in \mathcal{E}$ for any $v_i, v_j \in \mathcal{V}$ is said to be undirected, where the edge (v_i, v_j) denotes that agents v_i and v_j can obtain information from each other (see Fig. 1.1 (b)).

For an edge (v_i, v_j), node v_i is called the parent node, v_j is the child node, and v_i is a neighbour of v_j . The set of neighbours of node v_i is denoted as $\mathcal{N}_i$, whose cardinality is called the in-degree of node v_i. A directed graph is strongly connected if there is a directed path from every node to every other node (see Fig. Fig. 1.1 (c)). Note that for an undirected graph, strong connectedness is simply termed connectedness. A graph is defined as being balanced when it has the same number of ingoing and outgoing edges for all the nodes (see Fig. 1.1 (d)). Clearly, an undirected graph is a special balanced graph.

A directed path from node v_{i_1} to node v_{i_l} is a sequence of ordered edges of the form $(v_{i_k}, v_{i_{k+1}}), k = 1, 2, \cdots, l-1$. An undirected path in an undirected graph is defined analogously. A cycle is a directed path that starts and ends at the same node. A directed graph is complete if there is an edge from every node to every other node. A undirected tree is an undirected graph where all the nodes can be connected by the way of a single undirected path. A directed graph that contains a spanning tree is that there exists a node called the root, and this root has a directed path to every other node of the graph. (Fig. 1.2(a), for example). For undirected graphs, the existence of a directed spanning tree is equivalent to being connected. However, in directed graphs, the existence of a directed spanning tree is a weaker condition than being strongly connected. A strongly connected graph contains more than one directed spanning tree. For example, Figs. 1.2 show four different spanning trees.

Graph Matrices

Associated with the communication graph is its adjacency matrix $\mathcal{A} = [a_{ij}] \in \mathbb{R}^{N\times N}$, where the element a_{ij} denotes the connection between the agent i and agent j. $a_{ij} = 1$ if $(j, i) \in \mathcal{E}$, otherwise is zero, and $a_{ii} = 0$ for all nodes with the assumption that there exists no self loop. The Laplacian matrix $\mathcal{L} = [l_{ij}] \in \mathbb{R}^{N\times N}$ is defined by $l_{ii} = \sum_{j=1}^{N} a_{ij}$ and $l_{ij} = -a_{ij}$ when $i \neq j$. For the undirected graph in Fig. 1.1 (b), the adjacency matrix is given by

$$\mathcal{A} = \begin{bmatrix} 0 & 1 & 0 & 1 & 0 \\ 1 & 0 & 1 & 0 & 1 \\ 0 & 1 & 0 & 1 & 0 \\ 1 & 0 & 1 & 0 & 1 \\ 0 & 1 & 0 & 1 & 0 \end{bmatrix},$$

and the resultant Laplacian matrix is obtained as

$$\mathcal{L} = \begin{bmatrix} 2 & -1 & 0 & -1 & 0 \\ -1 & 3 & -1 & 0 & -1 \\ 0 & -1 & 2 & -1 & 0 \\ -1 & 0 & -1 & 3 & -1 \\ 0 & -1 & 0 & -1 & 2 \end{bmatrix}.$$

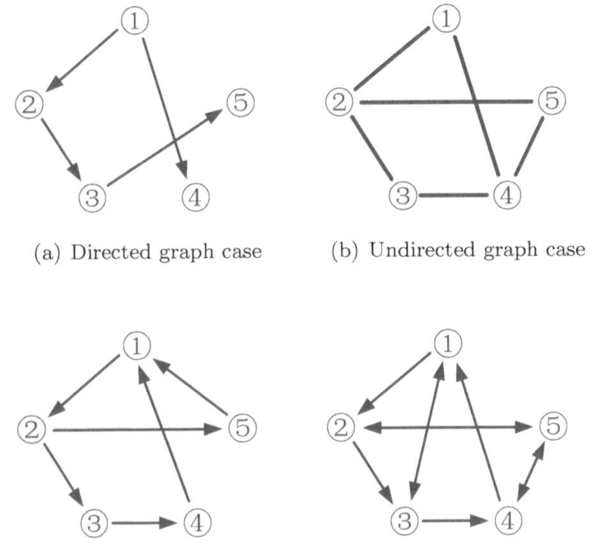

(a) Directed graph case (b) Undirected graph case

(c) Strongly connected graph case (d) Balanced and strongly connected case

FIGURE 1.1: The network connection illustration.

The eigenvalues of $\mathcal{L}$ are $\{0, 2, 2, 3, 5\}$. For the directed graph in Fig. 1.1 (c), the adjacency matrix is given by

$$\mathcal{A} = \begin{bmatrix} 0 & 0 & 0 & 1 & 1 \\ 1 & 0 & 0 & 0 & 0 \\ 0 & 1 & 0 & 0 & 0 \\ 0 & 0 & 1 & 0 & 0 \\ 0 & 1 & 0 & 0 & 0 \end{bmatrix},$$

and the resultant Laplacian matrix is obtained as

$$\mathcal{L} = \begin{bmatrix} 2 & 0 & 0 & -1 & -1 \\ -1 & 1 & 0 & 0 & 0 \\ 0 & -1 & 1 & 0 & 0 \\ 0 & 0 & -1 & 1 & 0 \\ 0 & -1 & 0 & 0 & 1 \end{bmatrix}.$$

The eigenvalues of $\mathcal{L}$ are $\{0, 1, 1.5 \pm j0.866, 2\}$.

From the definition of the Laplacian matrix and also the above examples, it is easy to see that $\mathcal{L}$ is diagonally dominant and has nonnegative diagonal entries. Since $\mathcal{L}$ has zero row sums, 0 is an eigenvalue of $\mathcal{L}$ with an associated eigenvector $\mathbf{1}$. According to Gershgorin's disc theorem, all nonzero eigenvalues

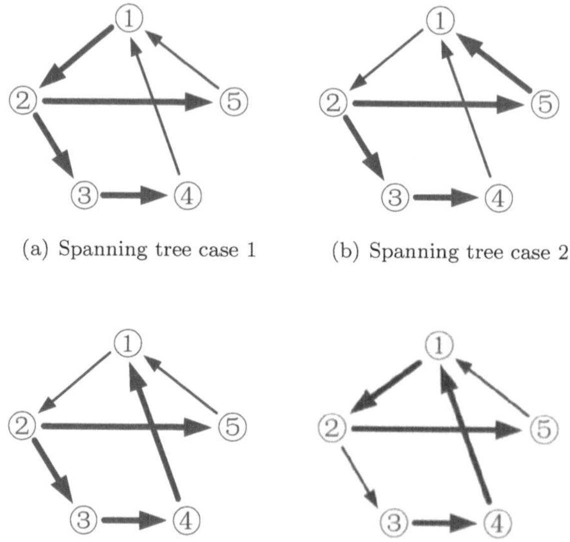

(a) Spanning tree case 1 (b) Spanning tree case 2

(c) Spanning tree case 3 (d) Spanning tree case 4

FIGURE 1.2: The network connection illustration.

of $\mathcal{L}$ are located within a disk in the complex plane centred at $d_{\max}$ and having radius of $d_{\max}$, where $d_{\max}$ denotes the maximum in-degree of all nodes. According to the definition of M-matrix in the last subsection, we know that the Laplacian matrix $\mathcal{L}$ is a singular M-matrix.

Properties

Some properties of the Laplacian matrix are given as follows.

Lemma 1.4.6 ([71, 125]) *For an undirected graph $\mathcal{G}$, if it is connected, the Laplacian matrix $\mathcal{L}$ has a single zero eigenvalue with $\mathcal{L}\mathbf{1}_N = 0$, and all the other eigenvalues of $\mathcal{L}$ are real and positive. Furthermore, we have the following properties for the Laplacian matrix $\mathcal{L}$:*

1. $x^T\mathcal{L}x = \frac{1}{2}\sum_{i=1}^{N}\sum_{j=1}^{N} a_{ij}(x_j - x_i)^2$;

2. $\lambda_2(\mathcal{L}) = \min_{x\neq 0, 1_N^T x=0} \frac{x^T\mathcal{L}x}{x^T x}$;

3. $x^T\mathcal{L}x \geq \lambda_2(\mathcal{L})x^T x$, *where $\lambda_2(\mathcal{L})$ is the smallest positive eigenvalue of $\mathcal{L}$.*

Lemma 1.4.7 *[127] If $\mathcal{G}(A)$ is strongly connected, then $\mathcal{L}$ has a simple eigenvalue at zero and there exists a positive vector $\delta = [\delta_1, \delta_2, \cdots, \delta_N]^T \in \mathbb{R}^N$ such that $\delta^T\mathcal{L} = 0$.*

Lemma 1.4.8 *[71, 125] The Laplacian matrix $\mathcal{L}$ of a directed graph $\mathcal{G}$ has at least one zero eigenvalue with a corresponding right eigenvector*

$\mathbf{1} = [1, 1, \ldots, 1]^T$ *and all nonzero eigenvalues have positive real parts. Furthermore, zero is a simple eigenvalue of* $\mathcal{L}$ *if and only if* $\mathcal{G}$ *has a directed spanning tree. In addition, there exists a nonnegative left eigenvector* r *of* $\mathcal{L}$ *associated with the zero eigenvalue, satisfying* $r^T\mathcal{L} = 0$ *and* $r^T 1 = 1$. *Moreover,* r *is unique if* $\mathcal{G}$ *has a directed spanning tree.*

Lemma 1.4.9 *[28] In the communication topology* $\mathcal{G}$ *with* N *followers and one leader indexed by 0, suppose that* N *followers contains a directed spanning tree with the leader as the root. Since the leader has no neighbours, the Laplacian matrix* $\mathcal{L}$ *has the following structure*

$$\mathcal{L} = \begin{bmatrix} 0 & 0_{1\times N} \\ \mathcal{L}_2 & \mathcal{L}_1 \end{bmatrix},$$

where $\mathcal{L}_1 \in \mathbb{R}^{N\times N}$ *and* $\mathcal{L}_2 \in \mathbb{R}^{N\times 1}$. *It can be seen that* $\mathcal{L}_1$ *is a nonsingular M-matrix and there exists a positive diagonal matrix* Q *such that*

$$Q\mathcal{L}_1 + \mathcal{L}_1^T Q \geq \rho_0 I, \tag{1.2}$$

for some positive constant ρ_0. Q *can be constructed by letting* $Q = \text{diag}\{q_1, q_2, \cdots, q_N\}$, *where* $q = [q_1, q_2, \cdots, q_N]^T = \left(\mathcal{L}_1^T\right)^{-1}[1, 1, \cdots, 1]^T$.

1.5 Notes

Some matarials in Sections 1.1 and 1.2 are from the first author's Ph.D. thesis [153]. The Definitions 4–11 and Theorem 1 related to stability theory in Section 1.4.3 is from [25]. The basic algebraic graph thoery in Section 1.4.4 is mainly from some monographs, such as [68, 71, 122, 127, 168].

Chapter 2

Stabilization of Single Systems with Input Delay: Prediction and Observation

This chapter first deals with control design for Lipschitz nonlinear systems with constant input delay. Based on a truncated prediction of the system state over the delay period, an output feedback control law is constructed. Within the framework of Lyapunov-Krasovskii functionals, a set of conditions are identified under which the closed-loop system is globally asymptotically stable at the origin. Then, the results are extended to time-varying input delay case. After that, the sensor fault is considered and a predictive descriptor observer approach is provided to solve the delay and fault problem with finite-dimensional controller design. Finally, numerical examples and experimental validation are included to demonstrate the effectiveness of the proposed designs.

2.1 Problem Formulation and Preliminaries

2.1.1 Problem Statement

We consider control design for a class of Lipschitz nonlinear systems with input delay

$$\begin{cases} \dot{x}(t) = Ax(t) + Bu(t-h) + \phi(x(t)), \\ y(t) = Cx(t) \end{cases}, \tag{2.1}$$

where $x \in \mathbb{R}^n$ is the state, $u \in \mathbb{R}^p$ is the input, $y \in \mathbb{R}^q$ is the measurement output, $h \in \mathbb{R}_+$ is the input delay, $A \in \mathbb{R}^{n\times n}$, $B \in \mathbb{R}^{n\times p}$ and $C \in \mathbb{R}^{q\times n}$ are constant matrices with (A, B) being controllable and (A, C) being observable, and $\phi : \mathbb{R}^n \to \mathbb{R}^n$, $\phi(0) = 0$, is a Lipschitz nonlinear function with a Lipschitz constant γ. For any two constant vectors $a, b \in \mathbb{R}^n$,

$$\|\phi(a) - \phi(b)\| \leq \gamma\|a - b\|. \tag{2.2}$$

The control objective in this chapter is to design the output feedback control such that the system (2.1) is globally asymptotically stable at the origin.

2.1.2 Predictor and Truncated Prediction

In this subsection, we first recall the well-known predictor-based feedback approach [3, 66]. Consider a linear input-delayed system

$$\dot{x}(t) = Ax(t) + Bu(t-h),$$

where $x \in \mathbb{R}^n$ denotes the state, $u \in \mathbb{R}^m$ denotes the control input, $A \in \mathbb{R}^{n\times n}$ and $B \in \mathbb{R}^{n\times m}$ are constant matrices, $h \in \mathbb{R}_+$ is input delay, which is known and constant. We design the feedback controller as

$$u(t) = Kx(t+h).$$

Then the finite-dimensional closed-loop system can be expressed as

$$\dot{x}(t) = (A+BK)x(t).$$

If we could find a possible control gain matrix K to stabilize the closed-loop system, then the control design problem is solved. However, it is unrealistic since the state information x at time $t+h$ cannot be obtained with direct measurement. A feasible way is to calculate the vector $x(t+h)$ as follows [66]:

$$\begin{aligned} x(t+h) &= e^{A(t+h)}x(0) + \int_0^{t+h} \mathrm{e}^{A(t+h-\tau)}Bu(\tau-h)\mathrm{d}\tau \\ &= e^{Ah}x(t) + \int_{t-h}^{t} \mathrm{e}^{A(t-\tau)}Bu(\tau)\mathrm{d}\tau. \end{aligned}$$

Therefore we can express the controller with an ideal predictor

$$u(t) = K\left(e^{Ah}x(t) + \int_{t-h}^{t} \mathrm{e}^{A(t-\tau)}Bu(\tau)\mathrm{d}\tau\right). \tag{2.3}$$

Compared with the standard feedback approach, the predictor-based control has the advantage that arbitrary long input delay can be compensated in the closed-loop system. The drawback of predictor-based feedback is that the control input ((2.3), for example) is infinite-dimensional and contains troublesome integral term. It may be unreachable sometimes due to the sensor restriction. To overcome this problem, in this part, we will introduce the truncated predictor feedback (TPF) method [188, 189, 198, 200]. Consider an input-delayed system

$$\dot{x}(t) = Ax(t) + Bu(t-h).$$

From the system dynamics, we have

$$x(t) = e^{Ah}x(t-h) + \int_{t-h}^{t} e^{A(t-\tau)}Bu(\tau-h)\mathrm{d}\tau.$$

The first term, $e^{Ah}x(t-h)$, is a truncated predictor of the state $x(t)$ based on $x(t-h)$. We take the control input as

$$u(t) = Ke^{Ah}x(t),$$

where $K \in \mathbb{R}^{n\times m}$ is a control gain matrix. The resultant closed-loop dynamics are given by

$$\dot{x}(t) = (A+BK)x(t) - BKd(t),$$

where

$$d(t) = \int_{t-h}^{t} e^{A(t-\tau)}Bu(\tau-h)d\tau.$$

In the TPF method, the troublesome integral term is ignored, and only the prediction based on the exponential of the systems matrix is used for control design. The controller is finite-dimensional and easy to implement since the integral of the input information is not needed.

2.1.3 Preliminary Results

Before moving into the main results, we present a couple of preliminary results which are useful for the stability analysis later. We first present a simplified version of a lemma in [199].

Lemma 2.1.1 *For a positive definite matrix P, the following identity holds*

$$e^{A^Tt}Pe^{At} - e^{\omega t}P = -e^{\omega t}\int_0^t e^{-\omega\tau}e^{A^T\tau}Re^{A\tau}d\tau, \tag{2.4}$$

where

$$R = -A^TP - PA + \omega P.$$

Furthermore, if R is positive definite,

$$e^{A^Tt}Pe^{At} < e^{\omega t}P. \tag{2.5}$$

Proof 1 *A direct evaluation gives*

$$\begin{aligned} e^{-\omega t}e^{A^Tt}Pe^{At} - P &= e^{(A-\frac{1}{2}\omega I)^Tt}Pe^{(A-\frac{1}{2}\omega I)t} - P \\ &= e^{(A-\frac{1}{2}\omega I)^Tt}Pe^{(A-\frac{1}{2}\omega I)t}\Big|_0^t \\ &= \int_0^t \frac{d}{d\tau}e^{(A-\frac{1}{2}\omega I)^T\tau}Pe^{(A-\frac{1}{2}\omega I)\tau} \\ &= -\int_0^t e^{(A-\frac{1}{2}\omega I)^T\tau}Re^{(A-\frac{1}{2}\omega I)\tau}d\tau. \end{aligned}$$

Multiplying the both sides of the above equation by $e^{\omega t}$ gives (2.4). The subsequent inequality (2.5) directly follows (2.4) when R is positive definite.

We next recall a lemma from [45].

Lemma 2.1.2 *For a positive definite matrix P, and a function $x : [a,b] \to \mathbb{R}^n$, with $a, b \in \mathbb{R}$ and $b > a$, the following inequality holds:*

$$\left(\int_a^b x^T(\tau)d\tau\right) P \left(\int_a^b x(\tau)d\tau\right) \leq (b-a)\int_a^b x^T(\tau)Px(\tau)d\tau. \tag{2.6}$$

A proof of the above result may be found in reference [45]. Since it is fairly straightforward, we show a direct verification of the inequality here.

Proof 2 *A direct evaluation gives*

$$\begin{aligned}
&\left(\int_a^b x^T(\tau)d\tau\right) P \left(\int_a^b x(\tau)d\tau\right)\\
&= \int_a^b \int_a^b x^T(s)Px(\tau)dsd\tau\\
&\leq \int_a^b \int_a^b \|P^{1/2}x(s)\|\|P^{1/2}x(\tau)\|dsd\tau\\
&\leq \int_a^b \int_a^b \frac{1}{2}(\|P^{1/2}x(s)\|^2 + \|P^{1/2}x(\tau)\|^2)dsd\tau\\
&= (b-a)\int_a^b \frac{1}{2}\|P^{1/2}x(s)\|^2 ds + (b-a)\int_a^b \frac{1}{2}\|P^{1/2}x(\tau)\|^2 d\tau\\
&= (b-a)\int_a^b \|P^{1/2}x(\tau)\|^2 d\tau\\
&= (b-a)\int_a^b x^T(\tau)Px(\tau)d\tau.
\end{aligned}$$

This completes the proof.

2.2 Truncated Prediction Feedback with Constant Input Delay

From the system dynamics (2.1), we have

$$x(t) = e^{Ah}x(t-h) + \int_{t-h}^t e^{A(t-\tau)}Bu(\tau-h)d\tau + \int_{t-h}^t e^{A(t-\tau)}\phi(x(\tau))d\tau. \tag{2.7}$$

The first term, $e^{Ah}x(t-h)$, is a truncated predictor of the system state at t based on $x(t-h)$. The truncated predictor output feedback control input is constructed as

$$\begin{aligned}\dot{\hat{x}}(t) &= A\hat{x}(t) + Bu(t-h) + \phi(\hat{x}(t)) + L(C\hat{x}(t) - y(t)), &(2.8)\\ u(t) &= Ke^{Ah}\hat{x}(t), &(2.9)\end{aligned}$$

where K and L denote the control gain matrix and the observer gain matrix, respectively, to be designed later. Let $\tilde{x}(t) = \hat{x}(t) - x(t)$. Substituting (2.9) into (2.1) and using (2.7) and (2.8) give the resultant closed-loop dynamics as

$$\begin{aligned}\dot{x}(t) &= Ax(t) + BKe^{Ah}\left(x(t-h) + \tilde{x}(t-h)\right) + \phi(x(t))\\ &= (A+BK)x(t) - BK(\lambda_1(t) + \lambda_2(t) + \lambda_3(t))\\ &\quad + BKe^{Ah}\tilde{x}(t-h) + \phi(x(t)), &(2.10)\\ \dot{\tilde{x}}(t) &= \dot{\hat{x}}(t) - \dot{x}(t) = (A+LC)\tilde{x}(t) + \phi(\hat{x}(t)) - \phi(x(t)), &(2.11)\end{aligned}$$

where

$$\lambda_1(t) = \int_{t-h}^{t} e^{A(t-\tau)}BKe^{Ah}x(\tau-h)\mathrm{d}\tau, \tag{2.12}$$

$$\lambda_2(t) = \int_{t-h}^{t} e^{A(t-\tau)}BKe^{Ah}\tilde{x}(\tau-h)\mathrm{d}\tau, \tag{2.13}$$

$$\lambda_3(t) = \int_{t-h}^{t} e^{A(t-\tau)}\phi(x(\tau))\mathrm{d}\tau. \tag{2.14}$$

It is worth mentioning that the extra terms λ_2 and λ_3 in (2.10) are due to the errors in the truncation, which correspond to the observation error and the nonlinear term $\phi(x(t))$, respectively.

Specifically, the controller and the observer gains in (2.8) and (2.9) are defined, respectively, as

$$K = -B^{\mathrm{T}}P_1 \quad \text{and} \quad L = -P_2^{-1}C^{\mathrm{T}}, \tag{2.15}$$

where P_1 and P_2 are both positive definite matrices to be designed.

Next, we establish the conditions for the positive definite matrices P_1 and P_2 such that the truncated predictor output feedback control in (2.8) and (2.9) stabilizes the system (2.1).

To start the analysis, let us try a Lyapunov function candidate

$$V_0 = x^{\mathrm{T}}(t)P_1x(t) + \tilde{x}^{\mathrm{T}}(t)P_2\tilde{x}(t). \tag{2.16}$$

With (2.15), the closed-loop dynamics in (2.10) and (2.11) can be rewritten as

$$\begin{aligned}\dot{x}(t) &= (A - BB^{\mathrm{T}}P_1)x(t) + BB^{\mathrm{T}}P_1(\lambda_1(t) + \lambda_2(t) + \lambda_3(t))\\ &\quad - BB^{\mathrm{T}}P_1e^{Ah}\tilde{x}(t-h) + \phi(x(t)), &(2.17)\\ \dot{\tilde{x}}(t) &= (A - P_2^{-1}C^{\mathrm{T}}C)\tilde{x}(t) + \phi(\hat{x}(t)) - \phi(x(t)). &(2.18)\end{aligned}$$

The derivative of V_0 along the trajectories of (2.17) and (2.18) can be evaluated as

$$\begin{aligned}\dot{V}_0 =& x^{\mathrm{T}}(t)(A^{\mathrm{T}}P_1 + P_1A - 2P_1BB^{\mathrm{T}}P_1)x(t) + \tilde{x}^{\mathrm{T}}(t)(A^{\mathrm{T}}P_2 + P_2A - 2C^{\mathrm{T}}C)\tilde{x}(t) \\ &+ 2x^{\mathrm{T}}(t)P_1BB^{\mathrm{T}}P_1(\lambda_1(t) + \lambda_2(t) + \lambda_3(t)) \\ &- 2x^{\mathrm{T}}(t)P_1BB^{\mathrm{T}}P_1e^{Ah}\tilde{x}(t-h) + 2x^{\mathrm{T}}(t)P_1\phi(x) + 2\tilde{x}(t)^{\mathrm{T}}P_2(\phi(\hat{x}) - \phi(x)) \\ \leq& x^{\mathrm{T}}(t)(A^{\mathrm{T}}P_1 + P_1A - P_1BB^{\mathrm{T}}P_1 + P_1P_1 + 3(P_1BB^{\mathrm{T}}P_1)^2)x(t) \\ &+ \Delta_1(t) + \Delta_2(t) + \Delta_3(t) + \tilde{x}^{\mathrm{T}}(t)(A^{\mathrm{T}}P_2 + P_2A - 2C^{\mathrm{T}}C + P_2P_2)\tilde{x}(t) \\ &+ \phi^{\mathrm{T}}(x)\phi(x) + (\phi(\hat{x}) - \phi(x))^{\mathrm{T}}(\phi(\hat{x}) - \phi(x)) \\ &+ \tilde{x}^{\mathrm{T}}(t-h)e^{A^{\mathrm{T}}h}P_1BB^{\mathrm{T}}P_1e^{Ah}\tilde{x}(t-h) \\ \leq& x^{\mathrm{T}}(t)(A^{\mathrm{T}}P_1 + P_1A - P_1BB^{\mathrm{T}}P_1 + (3\alpha^2+1)P_1P_1 + \gamma^2 I)x(t) \\ &+ \Delta_1(t) + \Delta_2(t) + \Delta_3(t) + \alpha e^{\omega_1 h}\tilde{x}^{\mathrm{T}}(t-h)P_1\tilde{x}(t-h) \\ &+ \tilde{x}^{\mathrm{T}}(t)(A^{\mathrm{T}}P_2 + P_2A - 2C^{\mathrm{T}}C + P_2P_2 + \gamma^2 I)\tilde{x}(t), \end{aligned} \tag{2.19}$$

where Lemma 2.1.1 is used to derive the last inequality provided that

$$R_1 = -A^{\mathrm{T}}P_1 - AP_1 + \omega_1 P_1 > 0, \tag{2.20}$$

with $\omega_1 \geq 0$, α is positive real numbers such that

$$\alpha I \geq P^{\frac{1}{2}}BB^{\mathrm{T}}P^{\frac{1}{2}}, \tag{2.21}$$

and

$$\Delta_i(t) = \lambda_i^{\mathrm{T}}(t)\lambda_i(t), \ i = 1, 2, 3.$$

The remaining part of the analysis is to explore the bounds on Δ_i, $i = 1, 2, 3$.

From (2.12) and (2.13), and by Lemmas 2.1.1 and 2.1.2 with the condition

$$R_2 = -A^{\mathrm{T}} - A + \omega_2 I > 0, \ \omega_2 \geq 0, \tag{2.22}$$

we have

$$\begin{aligned}\Delta_1 =& \int_{t-h}^{t} x^{\mathrm{T}}(\tau - h)e^{A^{\mathrm{T}}h}P_1BB^{\mathrm{T}}e^{A^{\mathrm{T}}(t-\tau)}\mathrm{d}\tau \\ &\times \int_{t-h}^{t} e^{A(t-\tau)}BB^{\mathrm{T}}P_1e^{Ah}x(\tau - h))\mathrm{d}\tau \\ \leq& h\int_{t-h}^{t} x^{\mathrm{T}}(\tau - h)e^{A^{\mathrm{T}}h}P_1BB^{\mathrm{T}}e^{A^{\mathrm{T}}(t-\tau)}e^{A(t-\tau)}BB^{\mathrm{T}}P_1e^{Ah}x(\tau - h))\mathrm{d}\tau \\ \leq& \alpha^2 h\int_{t-h}^{t} e^{\omega_2(t-\tau)}x^{\mathrm{T}}(\tau - h)e^{A^{\mathrm{T}}h}e^{Ah}x(\tau - h)\mathrm{d}\tau \\ \leq& \alpha^2 he^{2\omega_2 h}\int_{t-h}^{t} x^{\mathrm{T}}(\tau - h)x(\tau - h)\mathrm{d}\tau, \end{aligned} \tag{2.23}$$

and similarly,

$$\Delta_2 \leq \alpha^2 h e^{2\omega_2 h} \int_{t-h}^{t} \tilde{x}^{\mathrm{T}}(\tau - h)\tilde{x}(\tau - h)\mathrm{d}\tau, \tag{2.24}$$

where the inequality (2.21) is used.

On the other hand, from (2.2) and (2.14), we have

$$\begin{aligned}
\Delta_3 &= \int_{t-h}^{t} \phi^{\mathrm{T}}(x(\tau))e^{A^{\mathrm{T}}(t-\tau)}\mathrm{d}\tau \int_{t-h}^{t} e^{A(t-\tau)}\phi(x(\tau))\mathrm{d}\tau \\
&\leq h \int_{t-h}^{t} \phi^{\mathrm{T}}(x(\tau))e^{A^{\mathrm{T}}(t-\tau)}e^{A(t-\tau)}\phi(x(\tau))\mathrm{d}\tau \\
&\leq h \int_{t-h}^{t} e^{\omega_2(t-\tau)}\phi^{\mathrm{T}}(x(\tau))\phi(x(\tau))\mathrm{d}\tau \\
&\leq h e^{\omega_2 h}\gamma^2 \int_{t-h}^{t} x^{\mathrm{T}}(\tau)x(\tau)\mathrm{d}\tau.
\end{aligned} \tag{2.25}$$

For the term Δ_1 shown in (2.23), we consider the following Krasovskii functional

$$V_1(t) = e^h \int_{t-h}^{t} \left(e^{\tau-t}x^{\mathrm{T}}(\tau - h)x(\tau - h) + x^{\mathrm{T}}(\tau)x(\tau)\right)\mathrm{d}\tau.$$

A direct evaluation gives that

$$\begin{aligned}
\dot{V}_1(t) =& -e^h \int_{t-h}^{t} e^{\tau-t}x^{\mathrm{T}}(\tau - h)x(\tau - h)\mathrm{d}\tau \\
& - x^{\mathrm{T}}(t-2h)P_1x(t-2h) + e^h x^{\mathrm{T}}(t)x(t) \\
\leq& -\int_{t-h}^{t} x^{\mathrm{T}}(\tau - h)x(\tau - h)\mathrm{d}\tau + e^h x^{\mathrm{T}}(t)x(t).
\end{aligned} \tag{2.26}$$

For the term Δ_2 shown in (2.24), we consider the following Krasovskii functional

$$V_2(t) = e^h \int_{t-h}^{t} \left(e^{\tau-t}\tilde{x}^{\mathrm{T}}(\tau - h)\tilde{x}(\tau - h) + \tilde{x}^{\mathrm{T}}(\tau)\tilde{x}(\tau)\right)\mathrm{d}\tau.$$

A direct evaluation gives that

$$\dot{V}_2(t) \leq -\int_{t-h}^{t} \tilde{x}^{\mathrm{T}}(\tau - h)\tilde{x}(\tau - h)\mathrm{d}\tau + e^h\tilde{x}^{\mathrm{T}}(t)\tilde{x}(t). \tag{2.27}$$

For the term Δ_3 shown in (2.25), we consider the following Krasovskii functional

$$V_3(t) = e^h \int_{t-h}^{t} e^{\tau-t}x^{\mathrm{T}}(\tau)x(\tau)\mathrm{d}\tau$$

A direct evaluation gives that

$$\begin{aligned}\dot{V}_3(t) &= -e^h\int_{t-h}^{t} e^{\tau-t}x^{\mathrm{T}}(\tau)x(\tau)\mathrm{d}\tau + e^h x^{\mathrm{T}}(t)x(t) - x^{\mathrm{T}}(t-h)x(t-h)\\ &\le -\int_{t-h}^{t} x^{\mathrm{T}}(\tau)x(\tau)\mathrm{d}\tau + e^h x^{\mathrm{T}}(t)x(t). \end{aligned} \tag{2.28}$$

For the term $\tilde{x}^{\mathrm{T}}(t-h)P_1\tilde{x}(t-h)$ in (2.19), we consider the following Krasovskii functional

$$V_4(t) = \int_{t-h}^{t} \tilde{x}^{\mathrm{T}}(\tau)P_1\tilde{x}(\tau)\mathrm{d}\tau.$$

A direct evaluation gives that

$$\dot{V}_4(t) = \tilde{x}^{\mathrm{T}}(t)P_1\tilde{x}(t) - \tilde{x}^{\mathrm{T}}(t-h)P_1\tilde{x}(t-h). \tag{2.29}$$

Now, let

$$V = V_0 + \alpha^2 h e^{2\omega_2 h}(V_1 + V_2) + h e^{\omega_2 h}\gamma^2 V_3 + \alpha e^{\omega_1 h} V_4.$$

From (2.19) and (2.23)–(2.29), we have that

$$\dot{V} \le x^{\mathrm{T}}(t)Q_1 x(t) + \tilde{x}^{\mathrm{T}}(t)Q_2\tilde{x}(t), \tag{2.30}$$

where

$$\begin{aligned} Q_1 &= A^{\mathrm{T}}P_1 + P_1A - P_1BB^{\mathrm{T}}P_1 + (3\alpha^2+1)P_1P_1\\ &\quad + \alpha^2 h e^{(2\omega_2+1)h} I + \left(h e^{(\omega_2+1)h} + 1\right)\gamma^2 I, \end{aligned} \tag{2.31}$$

$$\begin{aligned} Q_2 &= A^{\mathrm{T}}P_2 + P_2A - 2C^{\mathrm{T}}C + P_2P_2 + \alpha e^{\omega_1 h}P_1\\ &\quad + \left(\gamma^2 + \alpha^2 h e^{(2\omega_2+1)h}\right) I. \end{aligned} \tag{2.32}$$

Theorem 2 *Consider the nonlinear system* (2.1). *The output feedback control law* (2.8) *and (2.9) with* (2.15) *globally asymptotically stabilizes the system* (2.1) *at the origin if the following conditions are satisfied, for* $W = P_1^{-1} > 0$, $P_2 > 0$, $\alpha > 0$, $\omega_1 \ge 0$, $\omega_2 \ge 0$,

$$\alpha W \ge BB^{\mathrm{T}}, \tag{2.33}$$

$$W\left(A - \frac{1}{2}\omega_1 I\right)^{\mathrm{T}} + \left(A - \frac{1}{2}\omega_1 I\right)W < 0, \tag{2.34}$$

$$\left(A - \frac{1}{2}\omega_2 I\right)^{\mathrm{T}} + \left(A - \frac{1}{2}\omega_2 I\right) < 0, \tag{2.35}$$

$$\begin{bmatrix} H_1 - BB^{\mathrm{T}} & W \\ W & -\frac{1}{\bar{\gamma}}I \end{bmatrix} < 0, \tag{2.36}$$

$$\begin{bmatrix} H_2 - 2C^{\mathrm{T}}C & P_2 \\ P_2 & -I \end{bmatrix} < 0, \tag{2.37}$$

where

$$\begin{aligned}\bar{\gamma} &= \alpha^2 h e^{(2\omega_2+1)h} + \left(h e^{(\omega_2+1)h} + 1\right)\gamma^2,\\ H_1 &= \left(3\alpha^2+1\right) I + W A^{\mathrm{T}} + AW,\\ H_2 &= \left(\gamma^2 + \alpha^2 h e^{(2\omega_2+1)h}\right) I + A^{\mathrm{T}} P_2 + P_2 A + \alpha e^{\omega_1 h} P_1.\end{aligned}$$

Proof 3 *It is easy to see that the conditions in* (2.33), (2.34) *and* (2.35) *are equivalent to the conditions specified in* (2.21), (2.20) *and* (2.22), *respectively. With* (2.31) *and* (2.32), *it can be shown by Schur Complement that conditions* (2.36) *and* (2.37) *are respectively equivalent to* $Q_1 < 0$ *and* $Q_2 < 0$, *which further implies from* (2.30) *that* $\dot{V}(t) < 0$. *Thus, the closed-loop dynamics* (2.10) *and* (2.11) *are globally asymptotically stable at the origin. This completes the proof.*

Remark 2.2.1 *The conditions shown in (2.33)–(2.37) can be checked by standard Linear Matrix Inequality routines for a set of fixed values. Note that* ω_2 *in* (2.35) *can be determined independently. The conditions in* (2.33), (2.34) *and* (2.36) *have to be checked simultaneously. With the possible solution* P_1 *obtained by computing* (2.33), (2.34) *and* (2.36), *a possible solution* P_2 *of* (2.37) *can then be computed for a fixed* P_1, *which indicates that the designs of the observer and the feedback law are coupled.*

2.3 Truncated Prediction Feedback with Time-Varying Input Delay

Consider a class of Lipschitz nonlinear systems in the presence of a time-varying input delay

$$\dot{x}(t) = Ax(t) + Bu(\phi(t)) + f(x(t)), \tag{2.38}$$

$$y(t) = Cx(t), \tag{2.39}$$

where $x \in \mathbb{R}^n$ is the state, $u \in \mathbb{R}^p$ is the input, $\phi(t) : \mathbb{R}_+ \to \mathbb{R}$ is a continuously differentiable function that incorporates the actuator delay, $A \in \mathbb{R}^{n\times n}$, $B \in \mathbb{R}^{n\times p}$ and $C \in \mathbb{R}^{q\times n}$ are constant matrices, and $f : \mathbb{R}^n \to \mathbb{R}^n$, with $f(0) = 0$, is a Lipschitz nonlinear function with a Lipschitz constant γ, i.e., for any two constant vectors $a, b \in \mathbb{R}^n$,

$$\|f(a) - f(b)\| \le \gamma\|a - b\|. \tag{2.40}$$

Note that the function $\phi(t)$ can be defined in a more intuitive form

$$\phi(t) = t - d(t), \tag{2.41}$$

where $d(t) \geq 0$ is the time-varying delay. However, it will be clear that the formalism (2.38) involving the function $\phi(t)$ facilitates the predictor design and the stability analysis where the inverse function $\phi^{-1}(t)$ is used [65]. In addition, a necessary assumption on $\phi(t)$ is needed for control design.

Assumption 2.3.1 *The function $\phi(t)$ is continuously differentiable, invertible and exactly known and is such that*

$$0 < \beta \leq \dot{\phi}(t) < \infty, \tag{2.42}$$

and there exists a finite number $h \geq 0$ such that, for all $t \geq 0$,

$$0 \leq d(t) \leq h, \tag{2.43}$$

i.e., the time-varying delay $d(t)$ is bounded.

Remark 2.3.1 *As pointed out in [200], the prediction time $\phi^{-1}(t) - t$ is also bounded since the delay $d(t)$ is bounded by* (2.43) *as in Assumption 2.3.1. In fact, we have*

$$0 \leq \phi^{-1}(t) - t \leq h. \tag{2.44}$$

The analytical solution of (2.38) can be computed as

$$x(t) = e^{A(t-\phi(t))}x(\phi(t)) + \int_{\phi(t)}^{t} e^{A(t-s)}Bu(\phi(s))\mathrm{d}s + \int_{\phi(t)}^{t} e^{A(t-s)}f(x(s))\mathrm{d}s. \tag{2.45}$$

The first term, $e^{A(t-\phi(t))}x(\phi(t))$, is a truncated predictor of the system state at t based on $x(\phi(t))$. In this section, we will utilize this truncated predictor in control design to globally asymptotically stabilize (2.38).

2.3.1 Stabilization by State Feedback

For system (2.38) with Assumption 2.3.1, we take the control structure as

$$u(t) = Ke^{A(\phi^{-1}(t)-t)}x(t) \tag{2.46}$$

where K is a control gain matrix to be specified later.

Remark 2.3.2 *It is noted that the control input* (2.46) *is build upon the existence of $\phi^{-1}(t)$ which requires the future knowledge of the delay. This may impose a limitation on some applications such as control over networks where the delay function $\phi(t)$ is generally discontinuous.*

Taking into account the solution (2.45), the closed-loop dynamics via the TPF control law (2.46) can be computed as

$$\begin{aligned} \dot{x}(t) &= Ax(t) + BKe^{A(t-\phi(t))}x(\phi(t)) + f(x(t)) \\ &= (A + BK)x(t) + f(x(t)) - BK(\lambda_1(t) + \lambda_2(t)), \end{aligned} \tag{2.47}$$

where

$$\lambda_1(t) = \int_{\phi(t)}^{t} e^{A(t-s)} BKe^{A(s-\phi(s))}x(\phi(s))\mathrm{d}s, \tag{2.48}$$

$$\lambda_2(t) = \int_{\phi(t)}^{t} e^{A(t-s)} f(x(s))\mathrm{d}s. \tag{2.49}$$

In (2.46), the control gain is specified as

$$K = -B^{\mathrm{T}}P_1, \tag{2.50}$$

where $P_1 = P_1^{\mathrm{T}} > 0$ is to be designed. Thus, the state feedback control design problem is to find a possible positive definite matrix P_1 such that the closed-loop system (2.47) is globally asymptotically stable at the origin. In other words, we will identify the conditions for P_1 under which the TPF control law (2.46) stabilizes the system.

Theorem 3 *Consider the Lipschitz nonlinear system* (2.38) *satisfying Assumption 2.3.1. The TPF control law* (2.46) *globally asymptotically stabilizes the system at the origin if there exist* $Y = P_1^{-1} > 0$, $\alpha > 0$ *and* $\omega_1 \geq 0$ *such that*

$$\alpha Y \geq BB^{\mathrm{T}}, \tag{2.51}$$

$$\left(A - \frac{1}{2}\omega_1 I\right)^{\mathrm{T}} + \left(A - \frac{1}{2}\omega_1 I\right) < 0, \tag{2.52}$$

$$\begin{bmatrix} YA^{\mathrm{T}} + AY - 2BB^{\mathrm{T}} + (2\alpha^2+1)I & Y \\ Y & -\frac{1}{\rho}I \end{bmatrix} < 0, \tag{2.53}$$

where $\rho = 2\alpha^2 h^2 e^{2\omega_1 h}\beta^{-1} + \left(1 + h^2 e^{\omega_1 h}\right)\gamma^2$.

Proof 4 *Consider a Lyapunov function candidate* $V_0(x(t)) = x^{\mathrm{T}}(t)P_1x(t)$. *The time derivative of* $V_0(x(t))$ *along the trajectories of* (2.47) *with* (2.50) *is*

$$\begin{aligned} \dot{V}_0 = &x^{\mathrm{T}}(t)\left(A^{\mathrm{T}}P_1 + P_1A - 2P_1BB^{\mathrm{T}}P_1\right)x(t) \\ &+ 2x^{\mathrm{T}}(t)P_1BB^{\mathrm{T}}P_1\left(\lambda_1(t) + \lambda_2(t)\right) + 2x^{\mathrm{T}}(t)P_1 f(x) \\ \leq &x^{\mathrm{T}}(t)\left(A^{\mathrm{T}}P_1 + P_1A - 2P_1BB^{\mathrm{T}}P_1 + P_1P_1 + 2(P_1BB^{\mathrm{T}}P_1)^2\right)x(t) \\ &+ \|\lambda_1(t)\|^2 + \|\lambda_2(t)\|^2 + \|f(x)\|^2 \\ \leq &x^{\mathrm{T}}(t)\left(A^{\mathrm{T}}P_1 + P_1A - 2P_1BB^{\mathrm{T}}P_1 + (2\alpha^2+1)P_1P_1 + \gamma^2 I\right)x(t) \\ &+ \|\lambda_1(t)\|^2 + \|\lambda_2(t)\|^2, \end{aligned} \tag{2.54}$$

where the Lipschitz condition (2.40) *is used and* α *is a positive real number such that*

$$\alpha I \geq P_1^{1/2}BB^{\mathrm{T}}P_1^{1/2}. \tag{2.55}$$

We next explore the bounds of $\|\lambda_1(t)\|^2$ *and* $\|\lambda_2(t)\|^2$ *used in the stability analysis.*

From (2.48) *and by using inequality* (2.6) *in Lemma 2.1.2, we have*

$$\begin{aligned}\|\lambda_1(t)\|^2 \leq & d(t)\int_{\phi(t)}^{t} x^{\mathrm{T}}(\phi(s))e^{A^{\mathrm{T}}(s-\phi(s))}P_1BB^{\mathrm{T}}e^{A^{\mathrm{T}}(t-s)} \\ & \times e^{A(t-s)}BB^{\mathrm{T}}P_1e^{A(s-\phi(s))}x(\phi(s))\mathrm{d}s \\ \leq & h\int_{\phi(t)}^{t} x^{\mathrm{T}}(\phi(s))e^{A^{\mathrm{T}}(s-\phi(s))}P_1BB^{\mathrm{T}}e^{A^{\mathrm{T}}(t-s)} \\ & \times e^{A(t-s)}BB^{\mathrm{T}}P_1e^{A(s-\phi(s))}x(\phi(s))\mathrm{d}s,\end{aligned}$$

where the bound (2.43) *has been used. By Lemma 2.1.1 with* $P = I$, *provided that*

$$R_1 := -A^{\mathrm{T}} - A + \omega_1 I > 0, \tag{2.56}$$

we have

$$\begin{aligned}\|\lambda_1(t)\|^2 \leq & \alpha^2 h\int_{\phi(t)}^{t} e^{\omega_1(t-s)}x^{\mathrm{T}}(\phi(s))e^{A(s-\phi(s))}e^{A(s-\phi(s))}x(\phi(s))\mathrm{d}s \\ \leq & \alpha^2 he^{2\omega_1 h}\int_{\phi(t)}^{t} x^{\mathrm{T}}(\phi(s))x(\phi(s))\mathrm{d}s.\end{aligned}$$

where we have inserted in the first inequality that

$$P_1(BB^T)(BB^T)P_1 \leq P_1(\alpha P_1^{-1})(\alpha P_1^{-1})P_1 = \alpha^2 I,$$

due to $BB^T \leq \alpha P_1^{-1}$ *which follows from* (2.55). *Introducing a change of variable* $\tau = \phi(s)$, *we have*

$$\mathrm{d}s = \left(\frac{\mathrm{d}}{\mathrm{d}s}\phi(s)\bigg|_{s=\phi^{-1}(\tau)}\right)^{-1}\mathrm{d}\tau. \tag{2.57}$$

Then, by Assumption 2.3.1, we have

$$\begin{aligned}\|\lambda_1(t)\|^2 \leq & \alpha^2 he^{2\omega_1 h}\int_{\phi(\phi(t))}^{\phi(t)} x^{\mathrm{T}}(\tau)x(\tau)\left(\frac{\mathrm{d}}{\mathrm{d}s}\phi(s)\bigg|_{s=\phi^{-1}(\tau)}\right)^{-1}\mathrm{d}\tau \\ \leq & \alpha^2 he^{2\omega_1 h}\beta^{-1}\int_{\phi(\phi(t))}^{\phi(t)} x^{\mathrm{T}}(\tau)x(\tau)\mathrm{d}\tau \\ \leq & \alpha^2 he^{2\omega_1 h}\beta^{-1}\int_{t-2h}^{t} x^{\mathrm{T}}(\tau)x(\tau)\mathrm{d}\tau.\end{aligned} \tag{2.58}$$

Similarly, by Lemma 2.1.2, we have from (2.49) *that*

$$\begin{aligned}\|\lambda_2(t)\|^2 &\leq h \int_{\phi(t)}^{t} f^{\mathrm{T}}(x(s)) e^{A^{\mathrm{T}}(t-s)} e^{A(t-s)} f(x(s)) \mathrm{d}s \\ &\leq h \int_{\phi(t)}^{t} e^{\omega_1 (t-s)} f^{\mathrm{T}}(x(s)) f(x(s)) \mathrm{d}s \\ &\leq h e^{\omega_1 h} \gamma^2 \int_{t-h}^{t} x^{\mathrm{T}}(s) x(s) \mathrm{d}s, \end{aligned} \tag{2.59}$$

where the Lipschitz condition (2.40) *has been used.*

For the terms $\|\lambda_1(t)\|^2$ *and* $\|\lambda_2(t)\|^2$ *shown in* (2.54), *we consider the following two Krasovskii functionals*

$$\begin{aligned}V_1(x_t) &= \alpha^2 h e^{2\omega_1 h} \beta^{-1} \int_0^{2h} \left(\int_{t-s}^{t} x^{\mathrm{T}}(\tau) x(\tau) \mathrm{d}\tau \right) \mathrm{d}s, \\ V_2(x_t) &= h e^{\omega_1 h} \gamma^2 \int_0^{h} \left(\int_{t-s}^{t} x^{\mathrm{T}}(\tau) x(\tau) \mathrm{d}\tau \right) \mathrm{d}s,\end{aligned}$$

where $x_t(\theta) = x(t+\theta)$, $\theta \in [-2h, 0]$. *It can be verified that*

$$\dot{V}_1(x_t) = 2\alpha^2 h^2 e^{2\omega_1 h} \beta^{-1} x^{\mathrm{T}}(t) x(t) - \alpha^2 h e^{2\omega_1 h} \beta^{-1} \int_{t-2h}^{t} x^{\mathrm{T}}(\tau) x(\tau) \mathrm{d}\tau, \tag{2.60}$$

$$\dot{V}_2(x_t) = h^2 e^{\omega_1 h} \gamma^2 x^{\mathrm{T}}(t) x(t) - h e^{\omega_1 h} \gamma^2 \int_{t-h}^{t} x^{\mathrm{T}}(\tau) x(\tau) \mathrm{d}\tau. \tag{2.61}$$

From (2.54), (2.60) *and* (2.61), *we can obtain that the time derivative of the Lyapunov-Krasovskii functional*

$$V(x_t) = V_0(x(t)) + V_1(x_t) + V_2(x_t) \tag{2.62}$$

along the trajectories of the closed-loop system (2.47) *verifies that*

$$\dot{V}(x_t) \leq x^{\mathrm{T}}(t) \Xi x(t), \tag{2.63}$$

where

$$\begin{aligned}\Xi :=& A^{\mathrm{T}} P_1 + P_1 A - 2 P_1 B B^{\mathrm{T}} P_1 + (2\alpha^2 + 1) P_1 P_1 \\ &+ \left(2\alpha^2 h^2 e^{2\omega_1 h} \beta^{-1} + (1 + h^2 e^{\omega_1 h}) \gamma^2\right) I.\end{aligned} \tag{2.64}$$

Thus, global asymptotic stability can be established by the Lyapunov-Krasovskii Stability Theorem [46] if the inequalities (2.55), (2.56) *and* $\Xi < 0$ *hold. Then, it is straightforward to verify that the inequalities* (2.55) *and* (2.56) *are equivalent to the conditions specified in* (2.51) *and* (2.52), *respectively. From* (2.64), *we can verify that, with* $Y = P_1^{-1}$, $\Xi < 0$ *is equivalent to*

$$\begin{aligned}&Y A^{\mathrm{T}} + A Y - 2 B B^{\mathrm{T}} + (2\alpha^2 + 1) I \\ &\quad + \left(2\alpha^2 h^2 e^{2\omega_1 h} \beta^{-1} + (1 + h^2 e^{\omega_1 h}) \gamma^2\right) Y Y < 0,\end{aligned} \tag{2.65}$$

which, in turn, is equivalent to (2.53). *This completes the proof.*

2.3.2 Stabilization by Output Feedback

In this subsection, it is assumed that only the system output (2.39) is measurable. We construct the following observer-based TPF control law:

$$\dot{\hat{x}}(t) = A\hat{x}(t) + Bu(\phi(t)) + L\left(C\hat{x}(t) - y(t)\right), \tag{2.66}$$

$$u(t) = Ke^{A(\phi^{-1}(t)-t)}\hat{x}(t), \tag{2.67}$$

where K and L are the control gain matrix and the observer gain matrix, respectively, to be specified soon.

In the following stability analysis, we will show that only the knowledge of the Lipschitz constant γ in (2.40) is required.

Let $\tilde{x}(t) = \hat{x}(t) - x(t)$ be the observation error. The closed-loop dynamics can be obtained as

$$\begin{aligned}\dot{x}(t) = &(A + BK)x(t) - BK\left(\lambda_1(t) + \lambda_2(t) + \lambda_3(t)\right)\\ &+ f(x(t)) + BKe^{A(t-\phi(t))}\tilde{x}(\phi(t)), \end{aligned} \tag{2.68}$$

$$\dot{\tilde{x}}(t) = (A + LC)\tilde{x}(t) - f(x(t)), \tag{2.69}$$

where $\lambda_1(t)$ and $\lambda_2(t)$ are defined in (2.48) and (2.49), respectively, and

$$\lambda_3(t) = \int_{\phi(t)}^{t} e^{A(t-s)}BKe^{A(s-\phi(s))}\tilde{x}(\phi(s))\mathrm{d}s. \tag{2.70}$$

Note that the extra term λ_3 in (2.68) is introduced by the observation error in the truncation.

The controller and the observer gains in (2.67) and (2.66) are specified, respectively, as

$$K = -B^{\mathrm{T}}P_1 \quad \text{and} \quad L = -P_2^{-1}C^{\mathrm{T}}, \tag{2.71}$$

where $P_1 = P_1^{\mathrm{T}} > 0$ and $P_2 = P_2^{\mathrm{T}} > 0$ are to be designed. Then, the output feedback control design problem is to find possible positive definite matrices P_1 and P_2 such that the observer-based TPF control law (2.66)–(2.67) globally asymptotically stabilizes system (2.38) at the origin.

Theorem 4 *Consider the system* (2.38)*–(2.39) satisfying Assumption 2.3.1. The observer-based TPF control law* (2.66)*–(2.67) with* (2.71) *globally asymptotically stabilizes the system at the origin if there exist* $Y = P_1^{-1} > 0$, $P_2 > 0$,

$\alpha > 0$, $\omega_1 \geq 0$ and $\omega_2 \geq 0$ such that

$$\alpha Y \geq BB^{\mathrm{T}}, \tag{2.72}$$

$$\left(A - \frac{1}{2}\omega_1 I\right)^{\mathrm{T}} + \left(A - \frac{1}{2}\omega_1 I\right) < 0, \tag{2.73}$$

$$Y\left(A - \frac{1}{2}\omega_2 I\right)^{\mathrm{T}} + \left(A - \frac{1}{2}\omega_2 I\right)Y < 0, \tag{2.74}$$

$$\begin{bmatrix} YA^{\mathrm{T}} + AY - BB^{\mathrm{T}} + (3\alpha^2+1)I & Y \\ Y & -\frac{1}{\varrho}I \end{bmatrix} < 0, \tag{2.75}$$

$$\begin{bmatrix} A^{\mathrm{T}}P_2 + P_2A - 2C^{\mathrm{T}}C + \alpha e^{\omega_2 h}P_1 + \delta I & P_2 \\ P_2 & -I \end{bmatrix} < 0, \tag{2.76}$$

where $\varrho = 2\alpha^2h^2e^{2\omega_1 h}\beta^{-1} + (2+h^2e^{\omega_1 h})\gamma^2$ and $\delta = 2\alpha^2h^2e^{2\omega_1 h}\beta^{-1}$.

Proof 5 *Consider a Lyapunov function candidate $V_0(x(t),\tilde{x}(t)) = x(t)^{\mathrm{T}}P_1x(t) + \tilde{x}(t)^{\mathrm{T}}P_2\tilde{x}(t)$. With (2.71), the time derivative of V_0 along the trajectories of (2.68) and (2.69) can be computed as*

$$\begin{aligned}
\dot{V}_0 =& x^{\mathrm{T}}(t)\left(A^{\mathrm{T}}P_1 + P_1A - 2P_1BB^{\mathrm{T}}P_1\right)x(t) \\
&+ 2x^{\mathrm{T}}(t)P_1BB^{\mathrm{T}}P_1\left(\lambda_1(t)+\lambda_2(t)+\lambda_3(t)\right) \\
&- 2x^{\mathrm{T}}(t)P_1BB^{\mathrm{T}}P_1e^{A(t-\phi(t))}\tilde{x}(\phi(t)) + 2x^{\mathrm{T}}(t)P_1f(x) \\
&+ \tilde{x}^{\mathrm{T}}(t)\left(A^{\mathrm{T}}P_2 + P_2A - 2C^{\mathrm{T}}C\right)\tilde{x}(t) - 2\tilde{x}^{\mathrm{T}}(t)P_2f(x) \\
\leq& x^{\mathrm{T}}(t)\left(A^{\mathrm{T}}P_1 + P_1A - P_1BB^{\mathrm{T}}P_1 + 3(P_1BB^{\mathrm{T}}P_1)^2 + P_1P_1\right)x(t) \\
&+ \|\lambda_1(t)\|^2 + \|\lambda_2(t)\|^2 + \|\lambda_3(t)\|^2 \\
&+ \tilde{x}^{\mathrm{T}}(\phi(t))e^{A^{\mathrm{T}}(t-\phi(t))}P_1BB^{\mathrm{T}}P_1e^{A(t-\phi(t))}\tilde{x}(\phi(t)) \\
&+ \tilde{x}^{\mathrm{T}}(t)(A^{\mathrm{T}}P_2 + P_2A - 2C^{\mathrm{T}}C + P_2P_2)\tilde{x}(t) + 2\|f(x)\|^2.
\end{aligned}$$

In view of Lemma 2.1.1, provided that

$$R_2 = -A^{\mathrm{T}}P_1 - AP_1 + \omega_2P_1 > 0, \tag{2.77}$$

with $\omega_2 \geq 0$, we have

$$\begin{aligned}
\dot{V}_0 \leq& x^{\mathrm{T}}(t)\left(A^{\mathrm{T}}P_1 + P_1A - P_1BB^{\mathrm{T}}P_1 + (3\alpha^2+1)P_1P_1 + 2\gamma^2I\right)x(t) \\
&+ \|\lambda_1(t)\|^2 + \|\lambda_2(t)\|^2 + \|\lambda_3(t)\|^2 + \alpha e^{\omega_2 h}\tilde{x}^{\mathrm{T}}(\phi(t))P_1\tilde{x}(\phi(t)) \\
&+ \tilde{x}^{\mathrm{T}}(t)\left(A^{\mathrm{T}}P_2 + P_2A - 2C^{\mathrm{T}}C + P_2P_2\right)\tilde{x}(t),
\end{aligned} \tag{2.78}$$

where α has been given in (2.55).

The remaining part is to explore the bound of $\|\lambda_3(t)\|^2$ since the bounds of $\|\lambda_1(t)\|^2$ and $\|\lambda_2(t)\|^2$ have been derived in (2.58) and (2.59). The same technique can be employed as in the derivation of (2.58) to compute the bound

$$\|\lambda_3(t)\|^2 \leq \alpha^2he^{2\omega_1 h}\beta^{-1}\int_{t-2h}^{t}\tilde{x}^{\mathrm{T}}(\tau)\tilde{x}(\tau)\mathrm{d}\tau. \tag{2.79}$$

For the terms $\|\lambda_3(t)\|^2$ *and* $\tilde{x}^{\mathrm{T}}(\phi(t))P_1\tilde{x}(\phi(t))$ *shown in* (2.78), *we consider another two Krasovskii functionals*

$$V_3 = \alpha^2 h e^{2\omega_1 h}\beta^{-1}\int_0^{2h}\left(\int_{t-s}^{t}\tilde{x}^{\mathrm{T}}(\tau)\tilde{x}(\tau)\mathrm{d}\tau\right)\mathrm{d}s,$$

$$V_4 = \alpha e^{\omega_2 h}\int_{\phi(t)}^{t}\tilde{x}^{\mathrm{T}}(s)P_1\tilde{x}(s)\mathrm{d}s,$$

which implies that

$$\dot{V}_3 = 2\alpha^2 h^2 e^{2\omega_1 h}\beta^{-1}\tilde{x}^{\mathrm{T}}(t)\tilde{x}(t) - \alpha^2 h e^{2\omega_1 h}\beta^{-1}\int_{t-2h}^{t}\tilde{x}^{\mathrm{T}}(\tau)\tilde{x}(\tau)\mathrm{d}\tau, \tag{2.80}$$

$$\dot{V}_4 \leq \alpha e^{\omega_2 h}\tilde{x}^{\mathrm{T}}(t)P_1\tilde{x}(t) - \alpha e^{\omega_2 h}\beta\tilde{x}^{\mathrm{T}}(\phi(t))P_1\tilde{x}(\phi(t)), \tag{2.81}$$

where the assumption (2.42) *on* $\dot{\phi}$ *has been used in* (2.81).

From (2.60), (2.61), (2.78), (2.80) *and* (2.81), *the time derivative of the Lyapunov-Krasovskii functional*

$$V(x_t, \tilde{x}_t) = V_0(x, \tilde{x}) + V_1(x_t) + V_2(x_t) + V_3(\tilde{x}_t) + V_4(\tilde{x}_t) \tag{2.82}$$

along the trajectories of the closed-loop dynamics (2.68) *and* (2.69) *verifies that*

$$\dot{W}(x_t, \tilde{x}_t) \leq x^{\mathrm{T}}(t)\Xi_1 x(t) + \tilde{x}^{\mathrm{T}}(t)\Xi_2\tilde{x}(t), \tag{2.83}$$

where

$$\begin{aligned}\Xi_1 &= A^{\mathrm{T}}P_1 + P_1A - P_1BB^{\mathrm{T}}P_1 + (3\alpha^2+1)P_1P_1\\ &\quad + \left(2\alpha^2h^2e^{2\omega_1 h}\beta^{-1} + (2 + h^2e^{\omega_1 h})\gamma^2\right)I, \qquad (2.84)\\ \Xi_2 &= A^{\mathrm{T}}P_2 + P_2A - 2C^{\mathrm{T}}C + P_2P_2 + \alpha e^{\omega_2 h}P_1 + 2\alpha^2h^2e^{2\omega_1 h}\beta^{-1}I. \qquad (2.85)\end{aligned}$$

Clearly, global asymptotic stability can be established by the Lyapunov-Krasovskii Stability Theorem [46] if the inequalities (2.55), (2.56), (2.77), $\Xi_1 < 0$ *and* $\Xi_2 < 0$ *hold simultaneously. With* $Y = P_1^{-1}$, *it is easy to see that these conditions are equivalent to* (2.72)–(2.76), *respectively. This completes the proof.*

Remark 2.3.3 *Compared with the conditions given in Theorem 3 for the state feedback stabilization, more stringent conditions are required in Theorem 4 for the output feedback case. If a possible solution* Y *is computed that verifies all conditions in Theorem 4, the solution* Y *also solves the state feedback stabilization problem described in Theorem 3.*

Remark 2.3.4 *With a possible solution* $P_1 = Y^{-1}$ *derived by computing the LMIs* (2.72)–(2.75), *a possible solution* P_2 *of* (2.76) *is then computed for the fixed* P_1. *This indicates that the designs of the observer and the feedback law are coupled. However, for large time delay and/or Lipschitz constant, a feasible solution* P_1 *tends to be very small, which implies from* (2.76) *that the observer design can be almost decoupled from the control design.*

2.4 Predictive Descriptor Observer Design for LTI Systems with Input Delay and Sensor Fault

2.4.1 Problem Formulation

We consider the stabilization problem of the following systems

$$\begin{cases} \dot{x}(t) = Ax(t) + Bu(t-h) \\ y(t) = Cx(t) + \omega(t) \end{cases}, \tag{2.86}$$

where $x \in \mathbb{R}^n$ is the state, $u \in \mathbb{R}^p$ is the input, $y \in \mathbb{R}^q$ is the output, $A \in \mathbb{R}^{n\times n}$, $B \in \mathbb{R}^{n\times p}$ and $C \in \mathbb{R}^{q\times n}$ are constant matrices with (A, B) being controllable and (A, C) being observable, h is a large constant delay, $\omega \in \mathbb{R}^q$ is the unknown but bounded sensor signal and satisfying

$$-\zeta \le \omega(t) \le \zeta, \quad t \ge 0,$$

with a bound $\zeta > 0$. As mentioned in [42], $\omega(t)$ can represent a combination of sensor disturbance and sensor fault. For example, $\omega(t)$ can be formulated as $\omega(t) = \mu(x, u) + f_s(t)$, where $\mu(x, u)$ represents sensor modeling uncertainties and $f_s(t)$ stands for possible sensor faults. For the sake of convenience, in this section we will use sensor fault for simplicity.

Next, we define a new state

$$x_1(t) = x(t + h).$$

A direct evaluation gives that

$$\dot{x}_1(t) = Ax(t + h) + Bu(t) = Ax_1(t) + Bu(t). \tag{2.87}$$

In this way, the first equation of the original system (2.86) is transformed into a delay-free system (2.87). If the state information $x(t)$ is measurable, we consider a controller

$$u(t) = Kx_1(t) = K\left[e^{Ah}x(t) + \int_t^{t+h} e^{A(t-s)}Bu(s-h)ds\right]. \tag{2.88}$$

From the definition of $x_1(t)$, we have

$$\|x(t)\| \le \|e^{-Ah}\|\|x_1(t)\| + h\|e^{-Ah}\|\left(\max_{-h\le\theta\le 0}\|e^{A\theta}\|\right)\|B\|\|K\|\|x_{1_t}(0)\|,$$

where $x_{1_t}(\theta) := x_1(t + \theta)$, $-h \le \theta \le 0$. Thus, $x(t) \to 0$ as $x_1(t) \to 0$. In other words, if $A + BK$ is Hurwitz, then the controller (2.88) stabilizes the transformed system (2.87) and the original system (2.86) is also stable with the same controller (2.88) [161].

Remark 2.4.1 *In the practical view, it is hard to directly implement the predictor feedback controller (2.88). First, more advanced actuators and processors are required to collect the historical information of the control input, which increases the cost and difficulty of the implementation. Second, the controller (2.88) is infinite-dimensional and contains distributed delays which make it impossible to obtain the exact control command in practice. Therefore, a sum of discrete delays are commonly used to replace the distributed delays. Unfortunately, this numerical approximation method cannot guarantee the system stability [37]. Third, the calculation of* $x_1(t)$ *requires accurate value of* $x(t)$*, which implies that this approach is not suitable for the case with sensor faults. To tackle these drawbacks, we will propose predictive descriptor observers to estimate the value of* $x_1(t)$ *and provide the corresponding controllers under both state and output feedback cases in the following part of this section.*

2.4.2 Stabilization by State Feedback

First, we consider the state information $x(t)$ is measurable. Then, the stabilization problem of system (2.86) is transformed into the stabilization problem of the following system:

$$\dot{x}(t) = Ax(t) + Bu(t-h). \tag{2.89}$$

Let $x_1(t) = x(t+h)$. A new system can be written as

$$\begin{cases} \dot{x}_1(t) = Ax_1(t) + Bu(t) \\ x(t) = x_1(t) + \omega_1(t) \end{cases}, \tag{2.90}$$

where $\omega_1(t) = x(t) - x_1(t) = x_1(t-h) - x_1(t)$.

Define a new state $\bar{x}_1(t) = [x_1^T(t), \omega_1^T(t)]^T \in \mathbb{R}^{2n}$. The state-space equation (2.90) can be rewritten in the augmented form

$$\begin{cases} E\dot{\bar{x}}_1(t) = \bar{A}\bar{x}_1(t) + \bar{B}u(t) + F\omega_1(t) \\ x(t) = \bar{C}\bar{x}_1(t) \end{cases}, \tag{2.91}$$

where

$$E = \begin{bmatrix} I_n & 0 \\ 0 & 0_{n\times n} \end{bmatrix}, \bar{A} = \begin{bmatrix} A & 0 \\ 0 & -I_n \end{bmatrix}, \bar{B} = \begin{bmatrix} B \\ 0 \end{bmatrix},$$
$$F = \begin{bmatrix} 0 & I_n \end{bmatrix}^T, \bar{C} = \begin{bmatrix} I_n & I_n \end{bmatrix}.$$

Motivated by descriptor observer design for systems with sensor faults in [42–44], a predictive descriptor observer is constructed as

$$\begin{cases} S\dot{\xi}(t) = (\bar{A} - L\bar{C})\,\xi(t) + \bar{B}u(t) - Fx(t) \\ \hat{\bar{x}}_1(t) = \xi(t) + S^{-1}Nx(t) \end{cases}, \tag{2.92}$$

where $\xi \in \mathbb{R}^{2n}$ is the state vector of the dynamic system, $\hat{\bar{x}}_1(t) = [\hat{x}_1^T(t), \hat{\omega}_1^T(t)]^T \in \mathbb{R}^{2n}$ is the estimation of $\bar{x}_1(t)$, L is the observer gain matrix to be designed later, and

$$S = \begin{bmatrix} I_n & 0 \\ M & M \end{bmatrix}, S^{-1} = \begin{bmatrix} I_n & 0 \\ -I_n & M^{-1} \end{bmatrix}, N = \begin{bmatrix} 0_{n\times n} \\ M \end{bmatrix}.$$

$M \in \mathbb{R}^{n\times n}$ is a nonsingular matrix providing more design degrees of freedom. Substituting $\xi(t) = \hat{\bar{x}}_1(t) - S^{-1}Nx(t)$ into the first equation of (2.92), we have that

$$\begin{aligned} S\dot{\hat{\bar{x}}}_1(t) &= \left(\bar{A} - L\bar{C}\right)\hat{\bar{x}}_1(t) - \bar{A}S^{-1}Nx(t) + L\bar{C}S^{-1}Nx(t) \\ &\quad + \bar{B}u(t) - Fx(t) + N\dot{x}(t) \\ &= \bar{A}\hat{\bar{x}}_1(t) + \bar{B}u(t) - L\bar{C}\hat{\bar{x}}_1(t) + L\bar{C}S^{-1}N\bar{C}\bar{x}_1(t) \\ &\quad - \bar{A}S^{-1}Nx(t) - Fx(t) + N\dot{x}(t) \\ &= \bar{A}\hat{\bar{x}}_1(t) + \bar{B}u(t) + L\bar{C}(\bar{x}_1(t) - \hat{\bar{x}}_1(t)) + N\dot{x}(t), \end{aligned}$$

where $\bar{A}S^{-1}N = -F$ and $\bar{C}S^{-1}N = I_n$ have been used for the derivation. Adding $N\dot{x}(t)$ to the first equation of (2.91) yields

$$S\dot{\bar{x}}_1(t) = \bar{A}\bar{x}_1(t) + \bar{B}u(t) + F\omega_1(t) + N\dot{x}(t). \tag{2.93}$$

where $E + N\bar{C} = S$ is used to obtain the above equation. The estimation error is defined by $e_1(t) = [e_{x_1}^T(t), e_{\omega_1}^T(t)]^T = \bar{x}_1(t) - \hat{\bar{x}}_1(t)$. Then, we obtain that

$$S\dot{e}_1(t) = \left(\bar{A} - L\bar{C}\right)e_1(t) + F\omega_1(t),$$

and

$$\dot{e}_1(t) = S^{-1}\left(\bar{A} - L\bar{C}\right)e_1(t) + d_1(t), \tag{2.94}$$

with $d_1(t) = FM^{-1}\omega_1(t)$. It is worth to mention that $S^{-1}F = FM^{-1}$ has been used for last derivation.

We propose a control design based on the estimated state information $\hat{x}_1(t)$. The control input takes the structure

$$u(t) = K\hat{x}_1(t), \tag{2.95}$$

where $K \in \mathbb{R}^{m\times n}$ is a gain matrix to be designed later. Under (2.95), the closed-loop system dynamics of (2.90) can be written as

$$\dot{x}_1(t) = Ax_1(t) + BK\hat{x}_1(t) = (A + BK)x_1(t) - BKe_{x_1}(t). \tag{2.96}$$

Theorem 5 *For system (2.90), the global asymptotic stability can be achieved by the predictive descriptor observer (2.92) and control algorithm (2.95) with*

$K = -B^T P_1$ and $L = SP_2^{-1}\bar{C}^T$ if there exist positive definite matrices P_1, P_2 and constants $\rho_0, \kappa_1, \kappa_2 > 0$ such that

$$\begin{bmatrix} WA^T + AW - (2-\kappa_1)BB^T & W \\ W & -\kappa_2\rho_0^{-1}/4 \end{bmatrix} < 0, \tag{2.97}$$

$$\begin{bmatrix} P_2 S^{-1}\bar{A} + \bar{A}^T S^{-T} P_2 - 2\bar{C}^T\bar{C} + \mu_1 & P_2 \\ P_2 & -\kappa_2^{-1} \end{bmatrix} < 0, \tag{2.98}$$

are satisfied with $W = P_1^{-1}$, $\mu_1 = [\kappa_1^{-1}P_1BB^TP_1 \quad 0; 0 \quad 0]$, where ρ_0 is a positive number such that $\rho_0 \geq \lambda_{max}(M^{-T}F^TFM^{-1})$ and $\lambda_{max}(\cdot)$ denotes the maximum eigenvalue of a matrix.

Proof 6 *To show the stability of the closed-loop system (2.96), a Lyapunov function candidate is defined as*

$$V(t) = x_1^T(t)P_1x_1(t) + e_1^T(t)P_2e_1(t) + 2\kappa_2^{-1}\rho_0 \int_t^{t+h} x_1^T(s-h)x_1(s-h)\mathrm{d}s,$$

where the parameters κ_2 and ρ_0 are defined in Theorem 5. The time derivative of $V(t)$ along the trajectory of (2.96) with $K = -B^TP_1$ and the trajectory of (2.94) with $L = SP_2^{-1}\bar{C}^T$ is

$$\begin{aligned} \dot{V} &= x_1^T(t)\left(A^* - 2P_1BB^TP_1\right)x_1(t) - 2x_1^T(t)P_1BKe_{x_1}(t) \\ &\quad - 2\kappa_2^{-1}\rho_0 x_1^T(t-h)x_1(t-h) + 2\kappa_2^{-1}\rho_0 x_1^T(t)x_1(t) \\ &\quad + e_1^T(t)\left(\bar{A}^* - 2\bar{C}^T\bar{C}\right)e_1(t) + 2e_1^T(t)P_2d_1(t), \end{aligned} \tag{2.99}$$

where $A^* = A^TP_1 + P_1A$ and $\bar{A}^* = P_2S^{-1}\bar{A} + \bar{A}^TS^{-T}P_2$. With Lemma 1.4.4, we have

$$-2x_1^T(t)P_1BKe_{x_1}(t) \leq \kappa_1 x_1^T(t)P_1BB^TP_1x_1(t) + \kappa_1^{-1}e_{x_1}^T(t)P_1BB^TP_1e_{x_1}(t),$$

and

$$2e_1^T(t)P_2d_1(t) \leq \kappa_2 e_1^T(t)P_2P_2e_1(t) + \kappa_2^{-1}\|d_1(t)\|^2.$$

where κ_1 is defined in Theorem 5. With $d_1(t) = FM^{-1}\omega_1(t)$ and the definition of $\omega_1(t)$ in (2.90), one further obtains that

$$\begin{aligned} & 2e_1^T(t)P_2d_1(t) \\ \leq\ & \kappa_2 e_1^T(t)P_2P_2e_1(t) + \kappa_2^{-1}\|FM^{-1}\omega_1(t)\|^2 \\ \leq\ & \kappa_2 e_1^T(t)P_2P_2e_1(t) + \kappa_2^{-1}\lambda_{max}(M^{-T}F^TFM^{-1})\|x_1(t-h) - x_1(t)\|^2 \\ \leq\ & \kappa_2 e_1^T(t)P_2P_2e_1(t) + 2\kappa_2^{-1}\rho_0 x_1^T(t)x_1(t) + 2\kappa_2^{-1}\rho_0 x_1^T(t-h)x_1(t-h). \end{aligned} \tag{2.100}$$

Then, putting the equations above together, it can be derived that

$$\dot{V}(t) \leq x_1^T(t)H_1x_1(t) + e_1^T(t)H_2e_1(t),$$

with

$$H_1 = A^* - (2-\kappa_1)P_1BB^TP_1 + 4\kappa_2^{-1}\rho_0 I_n, \tag{2.101}$$

$$H_2 = \bar{A}^* - 2\bar{C}^T\bar{C} + \kappa_2 P_2 P_2 + \mu_1, \tag{2.102}$$

where μ_1 is defined in Theorem 5. Clearly, global asymptotic stability can be established if $H_1 < 0$ and $H_2 < 0$ hold simultaneously. With $W = P_1^{-1}$, it can be seen that these conditions are equivalent to (2.97) and (2.98), respectively. This completes the proof.

Remark 2.4.2 *It is observed that, under the same condition, bigger h will leads to bigger upper bound of $\omega_1(t)$ in (2.94). Free nonsingular matrix M is added to adjust the upper bound of $d(t)$ and guarantee the observer performance.*

Remark 2.4.3 *Note that the state $x(t)$ is assumed to be measurable in this part, i.e., the sensor fault $\omega(t)$ in the original system (2.86) is not considered. The new state $x_1(t)$ and the disturbance $\omega_1(t)$ induced by time delay can be estimated from the predictive descriptor observer (2.92). On this basis, the feedback controller (2.95) does not contain the distributed delays in the traditional predictor feedback control and thus is easy to implement. Furthermore, the delay information is not needed during the controller/observer design and the stability anlaysis.*

2.4.3 Stabilization by Output Feedback under Sensor Fault

In this part, it is assumed that only the output information $y(t)$ in (2.86) is measurable with unknown but bounded sensor fault $\omega(t)$. We define $x_2(t) = x(t+h)$. Then, a new system can be written as

$$\begin{cases} \dot{x}_2(t) = Ax_2(t) + Bu(t) \\ y(t) = Cx_2(t) + \omega_2(t) \end{cases}, \tag{2.103}$$

where $\omega_2(t) = C(x_2(t-h) - x_2(t)) + \omega(t)$ represents the lumped disturbance induced by the time delay and sensor fault in the output channel. Define a new state $\bar{x}_2(t) = [x_2^T(t), \omega_2^T(t)]^T \in \mathbb{R}^{n+q}$. We have

$$\begin{cases} E_1\dot{\bar{x}}_2(t) = \bar{A}_1\bar{x}_2(t) + \bar{B}_1 u(t) + F_1\omega_2(t) \\ y(t) = \bar{C}_1\bar{x}_2(t) \end{cases},$$

where

$$E_1 = \begin{bmatrix} I_n & 0 \\ 0 & 0_{q\times q} \end{bmatrix}, \bar{A}_1 = \begin{bmatrix} A & 0 \\ 0 & -I_q \end{bmatrix}, \bar{B}_1 = \begin{bmatrix} B \\ 0 \end{bmatrix},$$

$$F_1 = \begin{bmatrix} 0 & I_q \end{bmatrix}^T, \ \bar{C}_1 = \begin{bmatrix} C & I_q \end{bmatrix}.$$

Similar to (2.92), a predictive descriptor observer is constructed as

$$\begin{cases} S_1\dot{z}(t) = \left(\bar{A}_1 - L_1\bar{C}_1\right) z(t) + \bar{B}_1 u(t) - F_1 y(t) \\ \hat{\bar{x}}_2(t) = z(t) + S_1^{-1} N_1 y(t) \end{cases}, \tag{2.104}$$

where $z(t) \in \mathbb{R}^{n+q}$ is the observer state, $\hat{\bar{x}}_2(t) = [\hat{x}_2^T(t), \hat{\omega}_2^T(t)]^T \in \mathbb{R}^{n+q}$ is the estimation of $\bar{x}_2(t)$, L_1 is the observer gain matrix to be designed later, and

$$S_1 = \begin{bmatrix} I_n & 0 \\ M_1 & M_1 \end{bmatrix}, S_1^{-1} = \begin{bmatrix} I_n & 0 \\ -I_n & M_1^{-1} \end{bmatrix}, N_1 = \begin{bmatrix} 0 \\ M_1 \end{bmatrix},$$

where $M_1 \in \mathbb{R}^{q \times q}$ is a nonsingular matrix providing more design degrees of freedom. The estimation error is defined by $e_2(t) = [e_{x_2}^T(t), e_{\omega_2}^T(t)]^T = \bar{x}_2(t) - \hat{\bar{x}}_2(t)$. Then, we obtain that

$$S_1\dot{e}_2(t) = \left(\bar{A}_1 - L_1\bar{C}_1\right) e_2(t) + F_1\omega_2(t),$$

and

$$\dot{e}_2(t) = S_1^{-1}\left(\bar{A}_1 - L_1\bar{C}_1\right) e_2(t) + d_2(t), \tag{2.105}$$

with $d_2(t) = F_1 M_1^{-1}\omega_2(t)$.

Lemma 2.4.1 *For the extra term $d_2(t)$ shown in (2.105), a bound can be established as*

$$\|d_2(t)\|^2 \leq 4\rho_1\rho_2 \|x_2(t-h)\|^2 + 4\rho_1\rho_2 \|x_2\|^2 + 2\rho_1\zeta^2,$$

where ρ_1 is a positive number such that $\rho_1 \geq \lambda_{max}\left(M_1^{-T}F_1^T F_1 M_1^{-1}\right)$ and ρ_2 is a positive number such that $\rho_2 \geq \lambda_{max}\left(C^T C\right)$.

Proof 7 *From the defination of $\omega_2(t)$ in (2.103), it can be derived that*

$$\begin{aligned} \|\omega_2(t)\|^2 &= \|C\left(x_2(t-h) - x_2(t)\right) + \omega(t)\|^2 \\ &\leq 2\|C(x_2(t-h) - x_2(t))\|^2 + 2\|\omega(t)\|^2 \\ &\leq 2\lambda_{max}\left(C^T C\right)\|(x_2(t-h) - x_2(t))\|^2 + 2\|\omega(t)\|^2 \\ &\leq 4\rho_2\|x_2(t-h)\|^2 + 4\rho_2\|x_2(t)\|^2 + 2\|\omega(t)\|^2. \end{aligned}$$

It follows that

$$\begin{aligned} \|d_2(t)\|^2 &= \left\|F_1 M_1^{-1}\omega_2(t)\right\|^2 \\ &\leq \lambda_{max}\left(M_1^{-T}F_1^T F_1 M_1^{-1}\right)\|\omega_2(t)\|^2 \\ &\leq 4\rho_1\rho_2\|x_2(t-h)\|^2 + 4\rho_1\rho_2\|x_2\|^2 + 2\rho_1\zeta^2. \end{aligned}$$

This completes the proof.

The control input takes the structure

$$u(t) = K_1\hat{x}_2(t). \tag{2.106}$$

Then, the closed-loop system dynamics of (2.103) can be written as

$$\dot{x}_2(t) = Ax_2(t) + BK_1\hat{x}_2(t) = (A + BK_1)x_2(t) - BK_1e_{x_2}(t). \tag{2.107}$$

Theorem 6 *For system (2.103), the state $x_2(t)$ and the observer error $e_2(t)$ are ultimately bounded under the observer (2.104) and the control algorithm (2.106) with $K_1 = -B^TP_3$ and $L_1 = S_1P_4^{-1}\bar{C}_1^T$ if there exist positive definite matrices P_3, P_4 and constants $\rho_1, \rho_2, \kappa_3, \kappa_4 > 0$ such that*

$$\begin{bmatrix} \bar{W}A^T + A\bar{W} - (2-\kappa_3)BB^T & \bar{W} \\ \bar{W} & -\kappa_4\rho_1^{-1}\rho_2^{-1}/8 \end{bmatrix} < 0, \tag{2.108}$$

$$\begin{bmatrix} P_4S_1^{-1}\bar{A}_1 + \bar{A}_1^TS_1^{-T}P_4 - 2\bar{C}_1^T\bar{C}_1 + \mu_2 & P_4 \\ P_4 & -\kappa_4^{-1} \end{bmatrix} < 0, \tag{2.109}$$

are satisfied with $\bar{W} = P_3^{-1}$, $\mu_2 = [\kappa_3^{-1}P_3BB^TP_3 \quad 0; 0 \quad 0]$, where $\rho_1 \geq \lambda_{max}\left(M_1^{-T}F_1^TF_1M_1^{-1}\right)$ and $\rho_2 \geq \lambda_{max}\left(C^TC\right)$ are defined in Lemma 2.4.1.

Proof 8 *Consider a Lyapunov function candidate as*

$$\bar{V}(t) = x_2^T(t)P_3x_2(t) + e_2^T(t)P_4e_2(t) + 4\kappa_4^{-1}\rho_1\rho_2\int_t^{t+h} x_2^T(s-h)x_2(s-h)\mathrm{d}s,$$

where the parameters κ_4, ρ_1 and ρ_2 are defined in Theorem 6. The time derivative of $\bar{V}(t)$ along the trajectory of (2.107) with $K_1 = -B^TP_3$ and the trajectory of (2.105) with $L = S_1P_4^{-1}\bar{C}_1^T$ is

$$\begin{aligned} \dot{\bar{V}}(t) &= x_2^T(t)\left(A_1^* - 2P_3BB^TP_3\right)x_2(t) - 2x_2^T(t)P_3BK_1e_{x_2}(t) \\ &\quad + e_2^T(t)\left(\bar{A}_1^* - 2\bar{C}_1^T\bar{C}_1\right)e_2(t) + 2e_2^T(t)P_4d_2(t) \\ &\quad - 4\kappa_4^{-1}\rho_1\rho_2x_2^T(t-h)x_2(t-h) + 4\kappa_4^{-1}\rho_1\rho_2x_2^T(t)x_2(t), \end{aligned} \tag{2.110}$$

where $A_1^* = A^TP_3 + P_3A$ and $\bar{A}_1^* = P_4S_1^{-1}\bar{A}_1 + \bar{A}_1^TS_1^{-T}P_4$. By using Lemma 1.4.4, one can obtain

$$-2x_2^T(t)P_3BK_1e_{x_2}(t) \leq \kappa_3x_2^T(t)P_3BB^TP_3x_2(t) + \kappa_3^{-1}e_{x_2}^T(t)P_3BB^TP_3e_{x_2}(t), \tag{2.111}$$

where κ_3 is a positive constant, and

$$\begin{aligned} 2e_2^T(t)P_4d_2(t) &\leq \kappa_4e_2^T(t)P_4P_4e_2(t) + \kappa_4^{-1}\|d_2(t)\|^2 \\ &\leq \kappa_4e_2^T(t)P_4P_4e_2(t) + 4\kappa_4^{-1}\rho_1\rho_2x_2^T(t-h)x_2(t-h) \\ &\quad + 4\kappa_4^{-1}\rho_1\rho_2x_2^T(t)x_2(t) + 2\kappa_4^{-1}\rho_1\zeta^2. \end{aligned} \tag{2.112}$$

Then, combining (2.110), (2.111) and (2.112), one can obtain

$$\dot{V}(t) \leq x_2^T(t)H_3x_2(t) + e_2^T(t)H_4e_2(t) + 2\kappa_4^{-1}\rho_1\zeta^2,$$

with

$$H_3 = A_1^* - (2-\kappa_3)P_3BB^TP_3 + 8\kappa_4^{-1}\rho_1\rho_2 I_n, \tag{2.113}$$
$$H_4 = \bar{A}_1^* - 2\bar{C}_1^T\bar{C}_1 + \kappa_4 P_4P_4 + \mu_2, \tag{2.114}$$

where μ_2 is defined in Theorem 6. It can be shown that conditions (2.108) and (2.109) are equivalent to $H_3 < 0$ and $H_4 < 0$, respectively, which further implies that $\dot{V}(t) < 2\kappa_4^{-1}\rho_1\zeta^2$. The positive term $2\kappa_4^{-1}\rho_1\zeta^2$ could be very small by choosing an appropriate free scaler κ_4 and nonsingluar matrix M_1. Therefore, we can conclude that the state $x_2(t)$ and the observation error $e_2(t)$ are ultimately bounded. This completes the proof.

Remark 2.4.4 *The existing results on descriptor observers are mostly about fault tolerant control for delay-free systems with sensor faults, such as [43, 44, 67, 90, 187]. Besides, the traditional predictor based approaches can only deal with input-delayed systems without sensor faults [65, 161] and are hard to be implemented. In this work, by using the predictive descriptor observer (2.104) and the feedback control input (2.106), the state of general LTI systems with large input delay and unknown sensor faults can converge to a small neighborhood of origin. Therefore, the result in this part are more applicable and favorable in practice.*

2.5 Numerical Examples

2.5.1 Constant Input Delay Case

We consider an input-delayed voltage feedback Chua's circuit system, as shown in Fig. 2.1. The new output feedback Chua's circuit [172] with input delay can be described by the following equations

$$\begin{cases} \frac{\mathrm{d}v_1}{\mathrm{d}t}(t) = \frac{G}{C_1}(v_2(t) - v_1(t)) - \frac{1}{C_1}f(v_1), \\ \frac{\mathrm{d}v_2}{\mathrm{d}t}(t) = \frac{1}{C_3}i_3(t) - \frac{G}{C_2}(v_2(t) - v_1(t)), \\ \frac{\mathrm{d}i_3}{\mathrm{d}t}(t) = -\frac{1}{L}(v_2(t) + R_0 i_3(t) + u(t-h)), \end{cases}$$

where v_1 and v_2 denote the voltage across C_1 and C_2, respectively, i_3 denotes the current through L, $G = 1/R$, and $f(v_1)$ characterizes the v-i property of the nonlinear resistor N_R with a slope G_a in the inner region and G_b in the outer region, and the breakpoint voltage E of the Chua's diode, i.e., $f(v_1) = G_b v_1 + 0.5(G_a - G_b)(|v_1 + E| - |v_1 - E|)$.

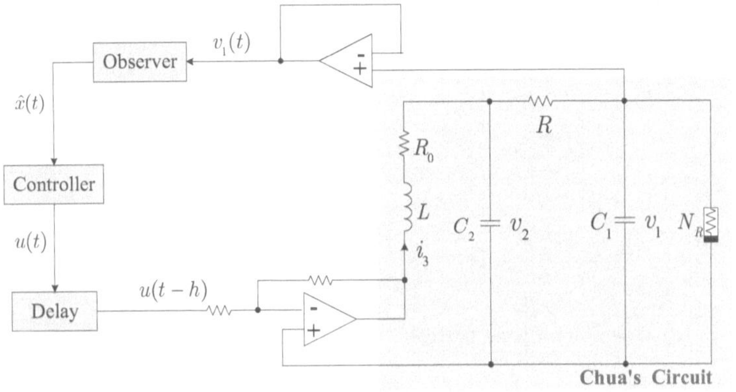

FIGURE 2.1: Output feedback Chua's circuit.

Let $x_1 = v_1/E$, $x_2 = v_2/E$, $x_3 = i_3/(EG)$, $\varepsilon = C_2/C_1$, $\rho = C_2/(LG^2)$, $r_0 = R_0C_2/(LG)$, $a = G_a/G$ $b = G_b/G$ and $\tau = Gt/C_2$. The following non-dimensional equations of the system can be formulated,

$$\begin{cases} \dot{x}_1 = \varepsilon(x_2 - (1+b)x_1 - \frac{a-b}{2}(|x_1+1| - |x_1-1|)), \\ \dot{x}_2 = x_1 - x_2 + x_3, \\ \dot{x}_3 = -\rho x_2 - r_0 x_3 - \rho u. \end{cases} \tag{2.115}$$

From Fig. 2.1, we have that the output is $y = v_1$. To facilitate the controller design, we re-arrange (2.115) into the state-space form of (2.1) with

$$A = \begin{bmatrix} -\varepsilon(1+b) & \varepsilon & 0 \\ 1 & -1 & 1 \\ 0 & -\rho & -r_0 \end{bmatrix}, \quad B = -\begin{bmatrix} 0 \\ 0 \\ \rho \end{bmatrix},$$

$$\phi = \begin{bmatrix} 0 \\ 0 \\ \frac{a-b}{2}(|x_1+1| - |x_1-1|) \end{bmatrix}, \quad C = \begin{bmatrix} E & 0 & 0 \end{bmatrix}.$$

For the simulation study, the following values of the circuit parameters are chosen:

$$C_1 = 10\,\text{nF},\ C_2 = 100\,\text{nF},\ L = 18.68\,\text{mH},\ R_0 = 16\,\Omega,$$
$$G_a = -0.75\,\text{mS},\ G_b = -0.41\,\text{mS},\ E = 1\,\text{V},\ R = 1.75\,\text{k}\Omega.$$

The circuit parameters imply that $\varepsilon = 10$, $\rho = 16.3945$, $r_0 = 0.1499$, $a = -1.3125$ and $b = -0.7175$. The time delay $h = 0.03\,\text{s}$ of the system is fixed, and the Lipschitz constant $\gamma = 5.95$ is computed.

With $\alpha = 0.01$ and $\omega_1 = \omega_2 = 0.1$, the solutions for P_1 and P_2 to satisfy (2.33)–(2.37) can be calculated as

$$P_1 = W^{-1} = \begin{bmatrix} 0.0045 & 0.0121 & 0.0042 \\ 0.0121 & 0.0703 & 0.0045 \\ 0.0042 & 0.0045 & 0.0064 \end{bmatrix},$$

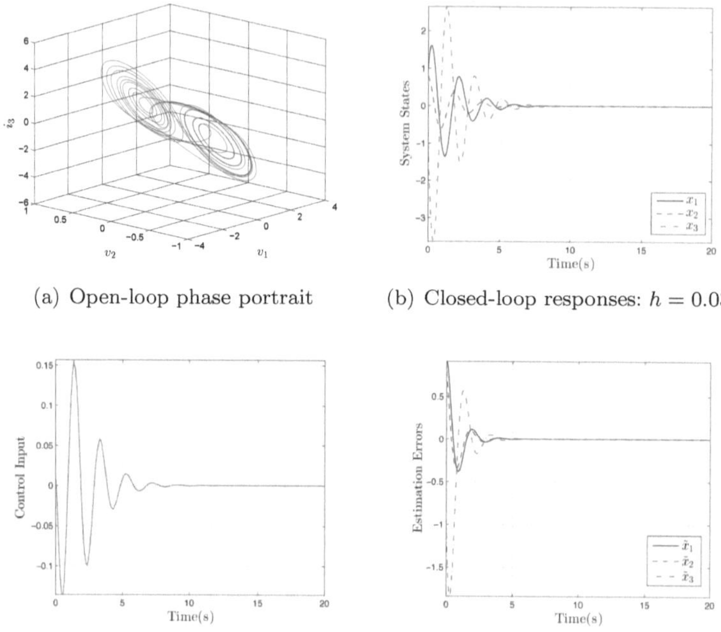

(a) Open-loop phase portrait (b) Closed-loop responses: $h = 0.03$

(c) Control input: $h = 0.03$ (d) Estimation errors: $h = 0.03$

FIGURE 2.2: Simulation results for canstant delay case.

and

$$P_2 = \begin{bmatrix} 0.5081 & -0.4567 & 0.2043 \\ -0.4567 & 1.4390 & -0.1414 \\ 0.2043 & -0.1414 & 0.1573 \end{bmatrix}.$$

By (2.15), we have

$$K = [0.0689, 0.0738, 0.1049],$$

and

$$L = \begin{bmatrix} -5.3798 \\ -1.1193 \\ 5.9824 \end{bmatrix}.$$

The initial conditions $x(0) = [1, 0.8, -0.9]^{\mathrm{T}}$ and $\hat{x}(0) = [0.1, 0, 0]^{\mathrm{T}}$ are chosen for the system (2.115) and the observer (2.8), respectively. The open-loop dynamic responses arc plotted in Fig. 2.2 (a), which shows the chaotic "double scroll" attractor pattern. With the inclusion of a voltage feedback loop into the nominal Chua's circuit, the asymptotic stability of the closed-loop system in the presence of input delay is achieved, as shown in Figs. 2.2 (b)–2.2 (d).

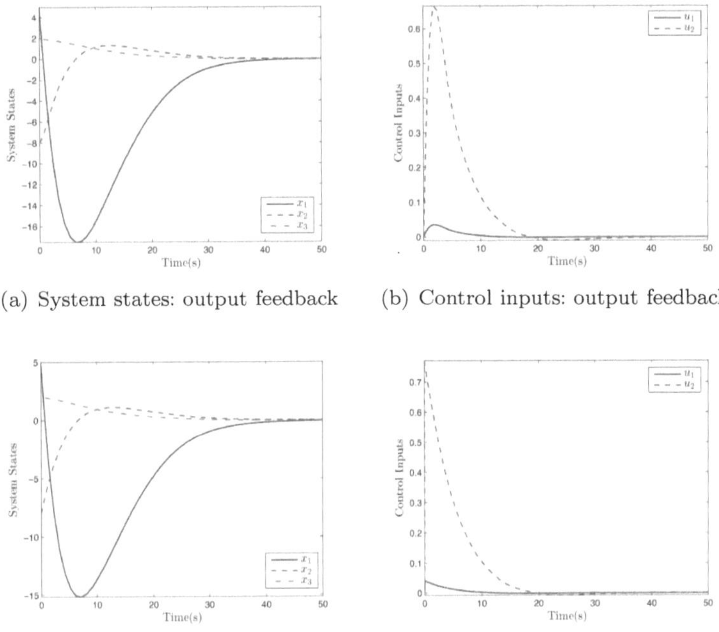

(a) System states: output feedback (b) Control inputs: output feedback

(c) System states: state feedback (d) Control inputs: state feedback

FIGURE 2.3: Simulation results for time-varying delay case.

2.5.2 Time-Varying Delay Case

We consider a third-order nonlinear system

$$\dot{x}(t) = \begin{bmatrix} 0.01 & 1 & 0 \\ 0 & -0.1 & 0.01 \\ 0 & -0.01 & -0.1 \end{bmatrix} x(t) + \begin{bmatrix} 0 & 0 \\ 0 & 2 \\ 2 & 0 \end{bmatrix} u(\phi(t)) + \begin{bmatrix} 0.005\sin(x_1(t)) \\ 0 \\ 0 \end{bmatrix}, \tag{2.116}$$

$$y(t) = \begin{bmatrix} 1 & 0 & 0 \\ 0 & 1 & 0 \end{bmatrix} x(t). \tag{2.117}$$

The linear part of this system represents an oscillator with a positive real pole, which is open-loop unstable, and the Lipschitz constant γ for the nonlinear function is 0.005. The time delay function $\phi(t)$ in the input is prescribed as

$$\phi(t) = t - 0.01\frac{t+1}{2t+1},$$

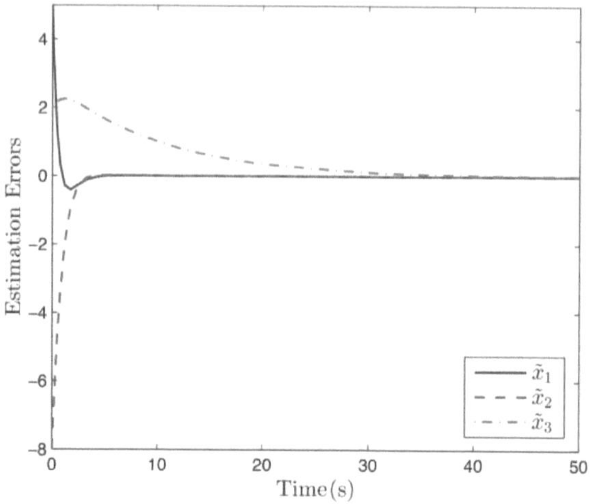

FIGURE 2.4: Observation errors: output feedback.

which implies that $h = 0.01$, $\beta = 1$ and its inverse function is

$$\phi^{-1}(t) = \frac{-2t - 0.99 + \sqrt{(0.99 - 2t)^2 + 8(t + 0.01)}}{4}.$$

The initial conditions $x(0) = [5, -8, 2]^{\mathrm{T}}$, $\hat{x}(0) = [0, 0, 0]^{\mathrm{T}}$ and $u(\theta) = 0$ for $\theta \in [\phi^{-1}(0), 0]$ are set in the simulation.

Let us design the observer-based TPF control law (2.66)–(2.67). With $\alpha = 0.15$ and $\omega_1 = \omega_2 = 1$, the solutions for $P_1 = Y^{-1}$ and P_2 to satisfy (2.72)–(2.76) can be computed as

$$P_1 = \begin{bmatrix} 0.0010 & 0.0053 & 0.0002 \\ 0.0053 & 0.0359 & 0.0020 \\ 0.0002 & 0.0020 & 0.0006 \end{bmatrix},$$

$$P_2 = \begin{bmatrix} 0.7644 & -0.2897 & -0.0005 \\ -0.2897 & 0.9781 & -0.0104 \\ -0.0005 & -0.0104 & 0.1222 \end{bmatrix},$$

and by (2.71), we then have

$$K = \begin{bmatrix} -0.0005 & -0.0039 & -0.0011 \\ -0.0106 & -0.0718 & -0.0039 \end{bmatrix}, L = \begin{bmatrix} -1.4738 & -0.4369 \\ -0.4369 & -1.1528 \\ -0.0426 & -0.0996 \end{bmatrix}.$$

Shown in Figs. 2.3 (a), 2.3 (b) and 2.4 are the simulation results of the closed-loop system under the output feedback control law. In these figures, we see that the convergence rate of the state estimation is much faster than the system response and the closed-loop performance is guaranteed. For comparison, if all states of the system (2.116) are measurable, the state feedback control law (2.46) is implemented. Without re-tuning any parameters, Figs. 2.3 (c), 2.3 (d) present the responses of the closed-loop system with the observer switched off. By a comparison, we see that in both cases the time evolutions of the system states (refer to Figs. 2.3 (a) and 2.3 (c)) are almost the same, while the control input plots are quite different before the observer recovers the system states.

2.6 Experiment Validation

In this section, a quadrotor tracking experiment is conducted to demonstrate the effectiveness of the proposed method in Section 2.4. To this end, we first give the quadrotor model and linearization treatment to obtain approximate LTI dynamics. Then, the experimental platform is introduced briefly. Finally, the experiment results and related discussions are given.

2.6.1 Quadrotor Model and Linearization

This subsection presents the linearized model of a quadrotor using the Euler Lagrange method [33, 169, 170]. As shown in Fig. 2.5, $O_eX_eY_eZ_e$ and $O_bX_bY_bZ_b$ denote the earth-fixed inertial frame and the body-fixed frame, respectively. The orientation of the rigid body is given by a rotation matrix $\boldsymbol{R}_e^b$:

$$\boldsymbol{R}_e^b = \begin{bmatrix} C\theta C\psi & C\theta S\psi & -S\theta \\ S\phi S\theta C\psi - S\psi C\phi & S\phi S\theta S\psi + C\phi C\psi & S\phi C\theta \\ C\phi S\theta C\psi + S\phi S\psi & C\phi S\theta S\psi - S\phi C\psi & C\phi C\theta \end{bmatrix},$$

where C and S are short for cosine and sine, respectively. Then, the dynamics of the quadrotor is described by

$$\begin{cases} \dot{\boldsymbol{p}}^e = \boldsymbol{v}^e \\ m\dot{\boldsymbol{v}}^e = -mg\boldsymbol{e}_z + \boldsymbol{R}_b^e \cdot T_b\boldsymbol{e}_z \\ \dot{\boldsymbol{\Theta}} = \boldsymbol{W} \cdot \boldsymbol{\omega}^b \\ \boldsymbol{I} \cdot \dot{\boldsymbol{\omega}}^b = \boldsymbol{\tau} - \boldsymbol{\omega}^b \times \boldsymbol{I}\boldsymbol{\omega}^b \end{cases}, \tag{2.118}$$

where $\boldsymbol{p}^e = \begin{bmatrix} x & y & z \end{bmatrix}$ denotes the position vector of the mass center of the quadrotor, $\boldsymbol{v}^e = \begin{bmatrix} v_x & v_y & v_z \end{bmatrix}$ denotes the velocity of the quadrotor in the inertial frame, $\boldsymbol{\omega}^b = \begin{bmatrix} \omega_x^b & \omega_y^b & \omega_z^b \end{bmatrix}^T$ denotes the angular velocity of the quadrotor in the body frame. m and $\boldsymbol{I}$ are the mass and the inertia matrix of the quadrotor, respectively. $\boldsymbol{\tau} = \begin{bmatrix} \tau_x & \tau_y & \tau_z \end{bmatrix}^T$ denotes the control torque vector, T_b denotes the thrust generated by rotors, $\boldsymbol{e}_z = \begin{bmatrix} 0 & 0 & 1 \end{bmatrix}^T$ is a unit

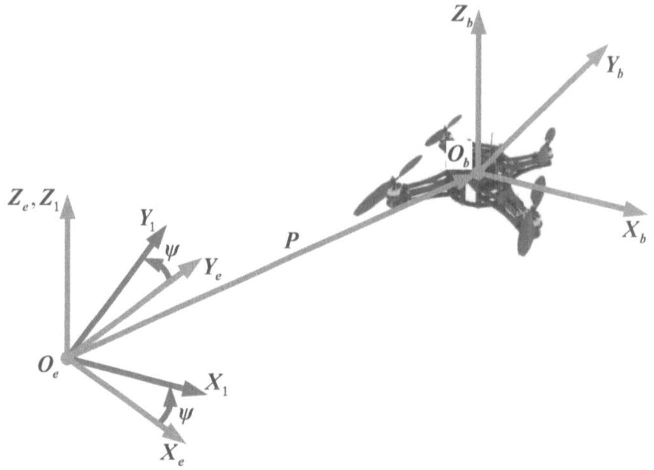

FIGURE 2.5: Coordinate system.

vector in the inertial frame, $\boldsymbol{\Theta} = \begin{bmatrix} \phi & \theta & \psi \end{bmatrix}^T$ denotes the Euler angle, $\boldsymbol{W}$ is a matrix defined as

$$\boldsymbol{W} = \begin{bmatrix} 1 & \sin\phi\tan\theta & \cos\phi\tan\theta \\ 0 & \cos\phi & \sin\phi \\ 0 & \sin\phi/\cos\theta & \cos\phi/\cos\theta \end{bmatrix}.$$

To simplify the tracking control design, the nonlinear model (2.118) will be transformed into a linear model with the following assumption.

Assumption 2.6.1 *The characteristic of the quadrotor has a small pitch angle θ and roll angle ϕ. The total pull force is approximately equal to its gravity. It can be expressed as* $\cos\theta \approx 1, \sin\theta \approx \theta, \cos\phi \approx 1, \sin\phi \approx \phi, T_b \approx mg$.

Note that the matrix $\boldsymbol{W}$ in (2.118) can be approximated as an identity matrix $\boldsymbol{I}_3$. Ignoring $-\boldsymbol{\omega}^b \times \boldsymbol{I}\boldsymbol{\omega}^b$, the model of quadrotor (2.118) can be simplified as

$$\begin{cases} \dot{\boldsymbol{p}}^e &= \boldsymbol{v}^e \\ \dot{\boldsymbol{v}}^e &= \boldsymbol{g}^e \\ \dot{\boldsymbol{\Theta}} &= \boldsymbol{\omega}^b \\ \boldsymbol{I}\cdot\dot{\boldsymbol{\omega}}^b &= \boldsymbol{\tau}. \end{cases} \tag{2.119}$$

where

$$\boldsymbol{g}^e = \begin{bmatrix} \dot{v}_x \\ \dot{v}_y \\ \dot{v}_z \end{bmatrix} = g\begin{bmatrix} \phi\sin\psi + \theta\cos\psi \\ -\phi\cos\psi + \theta\sin\psi \\ -1 \end{bmatrix} + \begin{bmatrix} 0 \\ 0 \\ \frac{T_b}{m} \end{bmatrix}.$$

Then the nonlinear model of the quadrotor is decoupled into three linear models:

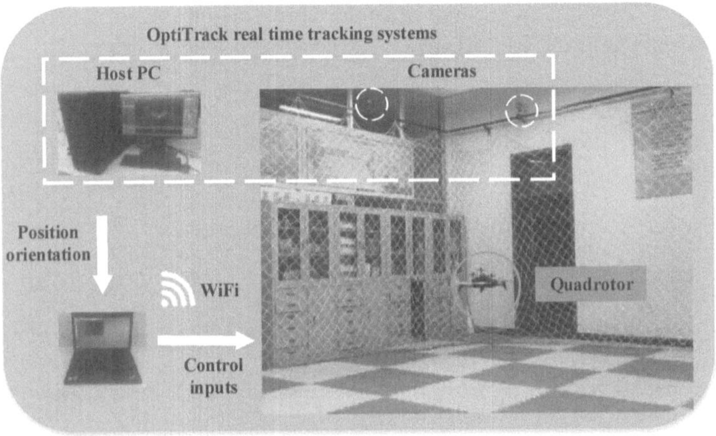

FIGURE 2.6: The quadrotor platform in System and Simulation Lab at Beijing Institute of Technology (BIT).

- Horizontal position channel model

$$\begin{cases} \dot{\boldsymbol{p}}_h = \boldsymbol{v}_h \\ \dot{\boldsymbol{v}}_h = u_h \end{cases}, \tag{2.120}$$

where $\boldsymbol{p}_h = \begin{bmatrix} x \\ y \end{bmatrix}$, $\boldsymbol{v}_h = \begin{bmatrix} v_x \\ v_y \end{bmatrix}$, and $u_h = -g\boldsymbol{A}_\psi \boldsymbol{\Theta}_h$ with $\boldsymbol{A}_\psi = \boldsymbol{R}_\psi \begin{bmatrix} 0 & 1 \\ -1 & 0 \end{bmatrix}$, $\boldsymbol{R}_\psi = \begin{bmatrix} \cos\psi & -\sin\psi \\ \sin\psi & \cos\psi \end{bmatrix}$ and $\boldsymbol{\Theta}_h = \begin{bmatrix} \phi \\ \theta \end{bmatrix}$.

- Height channel model

$$\begin{cases} \dot{p}_z = v_z \\ \dot{v}_z = u_z \end{cases}, \tag{2.121}$$

where $u_z = -g + \frac{T_b}{m}$.

- Attitude model

$$\begin{cases} \dot{\boldsymbol{\Theta}} &= \boldsymbol{\omega}^b \\ \boldsymbol{I} \cdot \dot{\boldsymbol{\omega}}^b &= \boldsymbol{\tau} \end{cases}. \tag{2.122}$$

Therefore, the control law of horizontal XY plane, height, and attitude can be designed separately.

To validate the proposed method by quadrotor tracking test, in the following experiment, a quadrotor is assumed to move in the XY plane and is designated to track a horizontal path. The height and the yaw angle of the quadrotor are controlled to be 1.2 m and 0 deg, respectively. From $u_h = -g\boldsymbol{A}_\psi \boldsymbol{\Theta}_h$, one can get the desired roll angle and pitch angle

$$\begin{bmatrix} \phi^d \\ \theta^d \end{bmatrix} = -\frac{1}{g} \cdot \boldsymbol{A}_\psi^{-1} \cdot u_h^d, \tag{2.123}$$

where u_h^d is the control input obtained from the proposed controller (2.106) for the approximate LTI system (2.120).

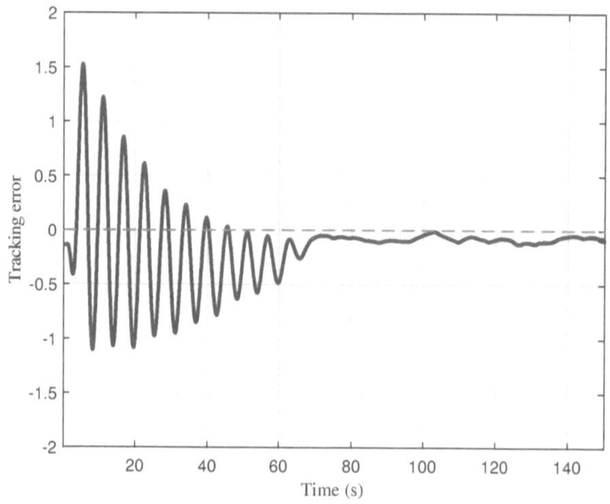

FIGURE 2.7: The tracking error.

2.6.2 Experimental Platform

An experimental platform is built to demonstrate the performance of the proposed method with real UAV systems. The platform is comprised of a Parrot Bebop quadrotor, the OptiTrack real-time tracking systems and a Linux-based computer, as shown in Fig. 2.6.

The real time tracking system is comprised of an array of eight OptiTrack Prime 13 cameras and a host PC. The function of the system is to identify the position and orientation of the quadrotor within the workspace area (5.5 m $\times$ 6 m $\times$ 2.5 m). The Linux-based computer receives these pose information and calculate the control input of the quadrotor according to the proposed method. The inputs are transmitted to Bebop quadrotors via 2.4 GHz Wi-Fi network. The robot operating systems (ROS) environment is used to transfer the messages.

2.6.3 Experimental Results

From (2.120), in this experiment, matrices A, B and C are given by

$$A = \begin{bmatrix} 0 & 1 \\ 0 & 0 \end{bmatrix}, \; B = \begin{bmatrix} 0 \\ 1 \end{bmatrix}, \; C = \begin{bmatrix} 1 & 0 \end{bmatrix}.$$

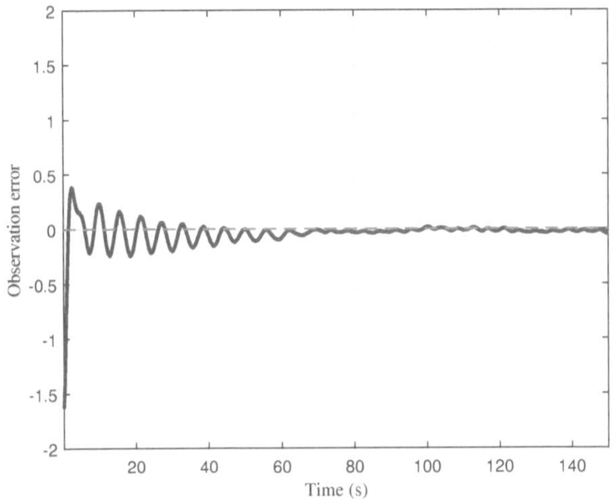

FIGURE 2.8: The observation error.

Limited by the experimental space, the input delay is set as $h = 1$ s. To characterize the input delay caused by remote communication in outdoor environment, in this experiment, we store the control input of the quadrotor generated at the current time t, and send the control input stored at $(t - h)$. In addition, the following fault signal is added into the measured position information:

$$\omega(t) = \begin{cases} 2t, & t \leq 10s, \\ 5, & 10s < t \leq 30s, \\ 5\sin(t-6), & t > 30s. \end{cases}$$

The initial states of quadrotors are chosen as $x_0 = [-1.5, 0, -1.5, 0.2]^T$. The proposed predictive descriptor observer (2.104) in Section 2.4 is applied to deal with the sensor fault and to compensate the delay effect. Figs. 2.7–2.9 show the tracking error, the observation error and the control input of the quadrotor, respectively.

It can be seen that the conditions specified in Theorem 6 can guarantee the tracking success of a quadrotor under sensor fault and long delay effect. It should be noted that due to the ultimate boundedness of the proposed method and the approximate linear model of the quadrotor, the final tracking and observation errors fluctuate around zero. Overall, the proposed predictive descriptor observer method is effective in the quadrotor tracking application and has potential in other applications such as cooperative control of multiple UAVs. Besides, it can be seen from the experimental results that it takes a long time for the controller to stabilize the system. One possible reason for the long time stabilization is that the stability analysis of this work is put in

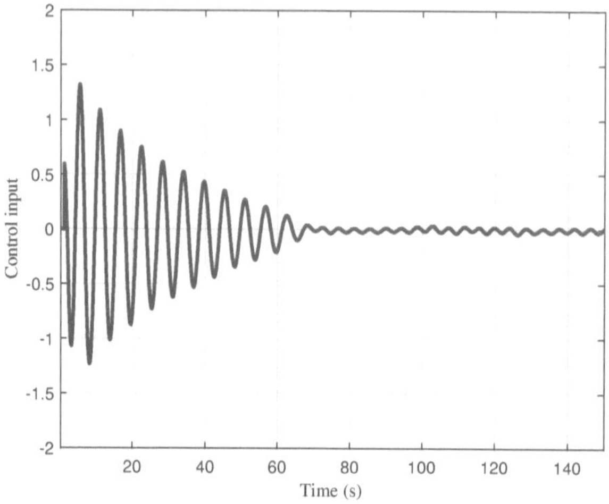

FIGURE 2.9: The control input.

the framework of Lyapunov-Krasovskii functionals due to the input delay and leads to certain conservatism in the presented conditions.

2.7 Conclusions

In this chapter, we have introduced the predictor feedback, the truncated prediction feedback and the predictive descriptor observer design for systems with constant or time-varying input delay. The stability analysises are carried out within the framework of Lyapunov-Krasovskii functionals in the time domain and sets of LMI conditions have been established for the stabilization of the input-delayed systems. Simulation results and experimental validation have shown the effectiveness of the proposed designs.

2.8 Notes

This chapter extends the ideas in [30, 42, 65, 66] and develops truncated prediction feedback and predictive descriptor observer design for a class of systems with constant or time-varying input delay. The main materials of this chapter are based on [156,206,207]. It is worth mentioning that in this chapter the input delays are assumed to be known. The results in this chapter can be extended to unknown input delay case by following the procedures shown in [159], and also can be extended to systems with parametric uncertainties by following the results in [211].

Chapter 3

Robust Consensus Control for Uncertian Linear Multi-Agent Systems with Input Delay

In this chapter, the consensus problem of uncertain multi-agent systems with input delay is investigated by exploiting the reduction method. First, the consensus analysis is derived based on real Jordan form of the Laplacian matrix. Then, further analysis has been developed to tackle the influence of the extra integral term under transformations due to the model uncertainty. Sufficient conditions are derived for the closed-loop system to achieve global consensus using Lyapunov-Krasovskii method in the time domain. Finally, the sumulation results prove the effectiveness of the proposed design.

3.1 Problem Formulation

In this chapter, we consider the control design for a set of N uncertain subsystems with input delay, of which the subsystems are described by

$$\dot{x}_i(t) = [A + \Delta A(t)]x_i(t) + [B + \Delta B(t)]u_i(t-h), \tag{3.1}$$

where, for subsystem i, $i = 1, 2, \ldots, N$, $x_i \in \mathbb{R}^n$ is the state vector, $u_i \in \mathbb{R}^m$ is the control input vector, $A \in \mathbb{R}^{n\times n}$ and $B \in \mathbb{R}^{n\times m}$ are constant matrices with (A, B) being controllable, $h > 0$ is the input delay, $x_i(\theta)$, $\theta \in [-h, 0]$, are given and bounded, $\Delta A(t)$ and $\Delta B(t)$ are time-varying uncertain matrices which can be formulated in the form [60] as

$$\Delta A(t) = E\Sigma(t)F_1 \quad \text{and} \quad \Delta B(t) = E\Sigma(t)F_2, \tag{3.2}$$

where E, F_1 and F_2 are real constant matrices with appropriate dimensions, and $\Sigma(t)$ is an unknown real time-varying matrix that satisfies $\Sigma^{\mathrm{T}}(t)\Sigma(t) \leq I$.

Remark 3.1.1 *It is worth noticing that the subsystems in the network are nominally identical and the model uncertainty matrices satisfy the same form as (3.2). Different from the existing works that focus on the identical agents in* *the network, the terms ΔA and ΔB in (1) allow the subsystems to have different dynamics and the uncertainty is characterised by the time-varying matrix $\Sigma(t)$, which implies that each subsystem in the group can be non-identical. For the consensus design, only the bound of $\Sigma(t)$ (i.e., the worst case) is needed.*

Assumption 3.1.1 *All subsystems in the network have known and identical input delays.*

Remark 3.1.2 *Assumption 3.1.1 is adopted for the convenience of illustration of the control design. The proposed method in this chapter may be extended to the network-connected systems with non-identical constant delays and even unknown time-varying delays if the worst case is taken into account in the design.*

Assumption 3.1.2 *0 is a simple eigenvalue of the Laplacian matrix.*

The consensus control problem considered in this chapter is to design a control strategy, using the relative state information, to ensure that all input-delayed uncertain subsystems converge to an identical trajectory.

Before moving into the consensus stability analysis, we next recall a lemma from [26] to reveal the block diagonal structure of the transformed Laplacian matrix for stability analysis.

Lemma 3.1.1 *For a Laplacian matrix $\mathcal{L}$ that satisfies Assumption 3.1.2, there exists a similarity transformation T, with its first column being $T_{(1)} = \mathbf{1}$, such that*

$$T^{-1}\mathcal{L}T = J, \tag{3.3}$$

with J being a block diagonal matrix in the real Jordan form

$$J = \begin{bmatrix} 0 & & & & & & \\ & J_2 & & & & & \\ & & \ddots & & & & \\ & & & J_p & & & \\ & & & & J_{p+1} & & \\ & & & & & \ddots & \\ & & & & & & J_q \end{bmatrix}_{N\times N}, \tag{3.4}$$

where $J_k \in \mathbb{R}^{n_k\times n_k}$, $k = 2, 3, \cdots, p$, are the Jordan blocks for real eigenvalues $\lambda_k > 0$ with the multiplicity n_k in the form

$$J_k = \begin{bmatrix} \lambda_k & 1 & & & \\ & \lambda_k & 1 & & \\ & & \ddots & \ddots & \\ & & & \lambda_k & 1 \\ & & & & \lambda_k \end{bmatrix}_{n_k\times n_k},$$

and $J_k \in \mathbb{R}^{2n_k \times 2n_k}$, $k = p+1, p+2, \dots, q$, *are the Jordan blocks for conjugate eigenvalues* $\alpha_k \pm j\beta_k$, $\alpha_k > 0$ *and* $\beta_k > 0$, *with the multiplicity* n_k *in the form*

$$J_k = \begin{bmatrix} \nu(\alpha_k, \beta_k) & I_2 & & & \\ & \nu(\alpha_k, \beta_k) & I_2 & & \\ & & \ddots & \ddots & \\ & & & \nu(\alpha_k, \beta_k) & I_2 \\ & & & & \nu(\alpha_k, \beta_k) \end{bmatrix}_{2n_k \times 2n_k},$$

with I_2 *the identity matrix in* $\mathbb{R}^{2\times 2}$ *and*

$$\nu(\alpha_k, \beta_k) = \begin{bmatrix} \alpha_i & \beta_i \\ -\beta_i & \alpha_i \end{bmatrix}_{2\times 2}.$$

3.2 Robust Consensus Controller Design

For the multi-agent systems (3.1), we consider the linear transformation

$$z_i(t) = x_i(t) + \int_t^{t+h} e^{A(t-\tau)} B u_i(\tau - h) d\tau, \tag{3.5}$$

by the reduction method. The original subsystems are transformed to

$$\begin{aligned} \dot{z}_i(t) =& (A + \Delta A) z_i(t) + D u_i(t) \\ &+ \Delta B u_i(t-h) - \Delta A \int_t^{t+h} e^{A(t-\tau)} B u_i(\tau - h) d\tau, \end{aligned} \tag{3.6}$$

where $D = e^{-Ah}B$. As seen in (3.6), system (3.1) is not completely reduced to a free-of-delay system due to the model uncertainties.

We propose a control design using the relative state information. The control input takes the structure

$$u_i(t) = -K \sum_{j=1}^{N} q_{ij}[z_i(t) - z_j(t)] = -K \sum_{j=1}^{N} l_{ij} z_j(t), \tag{3.7}$$

where $K \in \mathbb{R}^{m\times n}$ is a constant control gain matrix to be designed later.

Remark 3.2.1 *It is worth noting from* (3.5) *that the proposed control in* (3.7) *only uses the relative state information of the subsystems via network connections.*

Remark 3.2.2 *Note that the information on each control input* $u_i(t)$ *on the time interval* $[t-h, t]$ *can be stored and used for control. In practical implementations, the discretization of an integral or some numerical quadrature method [197] can be used to approximate the integral term in the control input* $u_i(t)$.

Remark 3.2.3 *For unknown non-identical time-varying delays $h_i(t)$ in (3.1) satisfying $0 < h_i(t) < \bar{h}$, $i = 1, 2, \ldots, N$, the upper bound of the delays that exist in all agents can be used in the predictor variable (3.5) as*

$$z_i(t) = x_i(t) + \int_t^{t+\bar{h}} e^{A(t-\tau)} B u_i(\tau - \bar{h}) \mathrm{d}\tau. \tag{3.8}$$

Then, the original subsystems (3.1) can be transformed to

$$\begin{aligned} \dot{z}_i(t) =& (A + \triangle A) z_i(t) + e^{-A\bar{h}} B u_i(t) - \triangle A \int_t^{t+\bar{h}} e^{A(t-\tau)} B u_i(\tau - \bar{h}) \mathrm{d}\tau \\ &+ \triangle B u_i(t - h_i(t)) + B\left(u_i(t - h_i(t)) - u_i(t - \bar{h})\right) \\ =& (A + \triangle A) z_i(t) + \overline{D} u_i(t) - \triangle A \\ &\times \int_t^{t+\bar{h}} e^{A(t-\tau)} B u_i(\tau - \bar{h}) \mathrm{d}\tau + \triangle B u_i(t - \bar{h}) \\ &- BK \sum_{j=1}^{N} l_{ij} \int_{t-\bar{h}}^{t-h_i(t)} \dot{z}_j(\tau) \mathrm{d}\tau - \triangle BK \sum_{j=1}^{N} l_{ij} \int_{t-\bar{h}}^{t-h_i(t)} \dot{z}_j(\tau) \mathrm{d}\tau, \end{aligned} \tag{3.9}$$

where $\overline{D} = e^{-A\bar{h}} B$. Compared with (3.6), the use of the upper-bound of delays leads to additional integral terms in (3.9), which further complicates the design. Fortunately, with the same control structure (3.7), our proposed design method is still applicable if K is carefully identified by introducing additional Krasovskii functionals in consensus analysis. Similar technique can be found in [120] for a single system with unknown time-varying delay but without parametric uncertainties.

Let $z(t) = [z_1, z_2, \ldots, z_N]^{\mathrm{T}}$, and the closed-loop system is then written as

$$\begin{aligned} \dot{z}(t) =& [I_N \otimes (A + \Delta A) - \mathcal{L} \otimes DK] z(t) \\ &- (\mathcal{L} \otimes \Delta BK) z(t-h) - (I_N \otimes \Delta A) \sigma(t), \end{aligned} \tag{3.10}$$

where $\sigma \triangleq [\sigma_1, \ldots, \sigma_N]^{\mathrm{T}}$ with the elements defined by

$$\sigma_i = -\int_t^{t+h} e^{A(t-\tau)} BK \sum_{j=1}^{N} l_{ij} z_j(\tau - h) \mathrm{d}\tau. \tag{3.11}$$

Define $r \in \mathbb{R}^N$ as the left eigenvector of $\mathcal{L}$ corresponding to the eigenvalue at 0, that is, $r^{\mathrm{T}} \mathcal{L} = 0$. Furthermore, let r be scaled such that $r^{\mathrm{T}} \mathbf{1} = 1$. It can be shown from Assumption 3.1.2 and Lemma 3.1.1 that there exists a non-singular matrix T with its first column being $T_{(1)} = \mathbf{1}$ and the first row of T^{-1} being $T_{(1)}^{-1} = r^{\mathrm{T}}$, such that

$$T^{-1} \mathcal{L} T = J. \tag{3.12}$$

Based on the vector r, we introduce a state transformation

$$\xi_i(t) = z_i(t) - \sum_{j=1}^{N} r_j z_j(t) \tag{3.13}$$

for $i = 1, \cdots, N$. With

$$\xi(t) = [\xi_1^{\mathrm{T}}(t), \xi_2^{\mathrm{T}}(t), \ldots, \xi_N^{\mathrm{T}}(t)]^{\mathrm{T}},$$

we have

$$\xi(t) = z(t) - [(\mathbf{1}r^{\mathrm{T}}) \otimes I_n]z(t) = (M \otimes I_n)z(t),$$

where $M \triangleq I_N - \mathbf{1}r^{\mathrm{T}}$. Since $r^{\mathrm{T}}\mathbf{1} = 1$, it can be shown that $M\mathbf{1} = 0$. Therefore, the consensus of system (3.10) is achieved when $\xi = 0$, as $\xi = 0$ implies $z_1 = z_2 = \cdots = z_N$, due to the fact the null space of M is span$\{\mathbf{1}\}$. Then, the consensus problem is now converted to the stabilization problem.

The dynamics of $\xi(t)$ can then be obtained as

$$\begin{aligned}\dot{\xi}(t) =& (M \otimes I_n)\dot{z}(t)\\ =& [I_N \otimes (A + \Delta A) - \mathcal{L} \otimes DK]\,\xi(t)\\ &- (\mathcal{L} \otimes \Delta BK)\xi(t-h) - (M \otimes I_n)(I_N \otimes \Delta A)\sigma. \end{aligned} \tag{3.14}$$

To explore the structure of $\mathcal{L}$ for stability analysis, let us introduce another state transformation

$$\eta(t) = (T^{-1} \otimes I_n)\xi(t). \tag{3.15}$$

Then, we have

$$\dot{\eta}(t) = [I_N \otimes (A + \Delta A) - J \otimes DK]\eta(t) - (J \otimes \Delta BK)\eta(t-h) - \Psi(z), \tag{3.16}$$

where $\Psi(z) = (T^{-1} \otimes I_n)(M \otimes I_n)(I_N \otimes \Delta A)\sigma$, $\eta(t) = [\eta_1(t), \eta_2(t), \ldots, \eta_N(t)]^{\mathrm{T}}$ and $\Psi(z) = [\psi_1(z), \psi_2(z), \ldots, \psi_N(z)]^{\mathrm{T}}$ with $\eta_i \in \mathbb{R}^n$ and $\psi_i : \mathbb{R}^{nN} \to \mathbb{R}^n$, for $i = 1, 2, \ldots, N$.

With the state transformations (3.13) and (3.15), we have

$$\eta_1(t) = (r^{\mathrm{T}} \otimes I_n)\xi(t) = [(r^{\mathrm{T}}M) \otimes I_n]z(t) \equiv 0. \tag{3.17}$$

The consensus control can be guaranteed by showing that η converges to zero, which is sufficed by showing that η_i converge to zero for $k = 2, 3, \ldots, N$, since we have shown that $\eta_1(t) \equiv 0$.

With the control law shown in (3.7), we design the control gain matrix K as

$$K = D^{\mathrm{T}}P, \tag{3.18}$$

where P is a positive definite matrix to satisfy certain condition. In the remaining part of the chapter, we will use Lyapunov-function-based analysis to identify conditions for P to ensure that the consensus control objective is achieved by using the control input (3.7) with the control gain (3.18).

The stability analysis will be carried out for η. Based on the structure of the Laplacian matrix shown in (3.4), we can see that

$$N_k = n_1 + \sum_{j=2}^{k} n_j, \tag{3.19}$$

with $n_1 = 1$ and $N_q = N$, where $k = 2, 3, \ldots, q$.

Due to the eigenstructure (3.4), the dynamics of the transformed state η in (3.16) will be discussed corresponding to the real and the complex eigenvalues, respectively.

For the state variables associated with the Jordan blocks J_k of real eigenvalues, i.e., for $2 \le k \le p$, we have the dynamics

$$\begin{aligned}\dot{\eta}_i(t) = &(A + \Delta A - \lambda_i DD^{\mathrm{T}}P)\eta_i(t) - DD^{\mathrm{T}}P\eta_{i+1}(t) - \lambda_i \Delta BD^{\mathrm{T}}P\eta_i(t-h)\\ &- \Delta BD^{\mathrm{T}}P\eta_{i+1}(t-h) - \psi_i(z), \quad i = N_{k-1}+1, N_{k-1}+2, \ldots, N_k - 1,\end{aligned} \tag{3.20}$$

$$\dot{\eta}_i(t) = (A + \Delta A - \lambda_i DD^{\mathrm{T}}P)\eta_i(t) - \lambda_i \Delta BD^{\mathrm{T}}P\eta_i(t-h) - \psi_i(z), \quad i = N_k. \tag{3.21}$$

For the state variables associated with the Jordan blocks J_k of conjugate complex eigenvalues, i.e., for $k > p$, we consider the dynamics of the state variables in pairs. For notational convenience, let

$$i_1(j) = N_{k-1} + 2j - 1 \quad \text{and} \quad i_2(j) = N_{k-1} + 2j, \tag{3.22}$$

where $j = 1, 2, \ldots, n_k/2$. The dynamics of $\eta_{i_1(j)}$ and $\eta_{i_2(j)}$, for $j = 1, 2, \ldots, n_k/2 - 1$, are expressed by

$$\begin{aligned}\dot{\eta}_{i_1(j)}(t) = &\left(A + \Delta A - \alpha_k DD^{\mathrm{T}}P\right)\eta_{i_1(j)}(t) - \beta_k DD^{\mathrm{T}}P\eta_{i_2(j)}(t)\\ &- DD^{\mathrm{T}}P\eta_{i_1(j)+2}(t) - \alpha_k \Delta BD^{\mathrm{T}}P\eta_{i_1(j)}(t-h)\\ &- \beta_k \Delta BD^{\mathrm{T}}P\eta_{i_2(j)}(t-h) - \Delta BD^{\mathrm{T}}P\eta_{i_1(j)+2}(t-h) - \psi_{i_1(j)}(z),\\ \dot{\eta}_{i_2(j)}(t) = &\left(A + \Delta A - \alpha_k DD^{\mathrm{T}}P\right)\eta_{i_2(j)}(t) + \beta_k DD^{\mathrm{T}}P\eta_{i_1(j)}(t)\\ &- DD^{\mathrm{T}}P\eta_{i_2(j)+2}(t) - \alpha_k \Delta BD^{\mathrm{T}}P\eta_{i_2(j)}(t-h)\\ &+ \beta_k \Delta BD^{\mathrm{T}}P\eta_{i_1(j)}(t-h) - \Delta BD^{\mathrm{T}}P\eta_{i_2(j)+2}(t-h) - \psi_{i_2(j)}(z),\end{aligned}$$

and, for $j = n_k/2$,

$$\begin{aligned}\dot{\eta}_{i_1(j)} = &\left(A + \Delta A - \alpha_k DD^{\mathrm{T}}P\right)\eta_{i_1(j)}(t) - \beta_k DD^{\mathrm{T}}P\eta_{i_2(j)}(t)\\ &- \alpha_k \Delta BD^{\mathrm{T}}P\eta_{i_1(j)}(t-h) - \beta_k \Delta BD^{\mathrm{T}}P\eta_{i_2(j)}(t-h) - \psi_{i_1(j)}(z),\\ \dot{\eta}_{i_2(j)} = &\left(A + \Delta A - \alpha_k DD^{\mathrm{T}}P\right)\eta_{i_2(j)}(t) + \beta_k DD^{\mathrm{T}}P\eta_{i_1(j)}(t)\\ &- \alpha_k \Delta BD^{\mathrm{T}}P\eta_{i_2(j)}(t-h) + \beta_k \Delta BD^{\mathrm{T}}P\eta_{i_1(j)}(t-h) - \psi_{i_2(j)}(z).\end{aligned}$$

Let

$$W_i = \eta_i^{\mathrm{T}}(t)P\eta_i(t). \tag{3.23}$$

For $i = N_{k-1}+1, N_{k-1}+2, \ldots, N_k - 1$, the time derivative of W_i along the trajectory (3.20) is

$$\begin{aligned}
\dot{W}_i =& \eta_i^{\mathrm{T}}(t)P\dot{\eta}_i(t) \\
=& \eta_i^{\mathrm{T}}(t)\left(A^{\mathrm{T}}P + PA - 2\lambda_k PDD^{\mathrm{T}}P\right)\eta_i(t) + 2\eta_i^{\mathrm{T}}(t)P\Delta A\eta_i(t) \\
& - 2\eta_i^{\mathrm{T}}(t)PDD^{T}P\eta_{i+1}(t) - 2\lambda_k\eta_i^{\mathrm{T}}(t)P\Delta BD^{\mathrm{T}}P\eta_i(t-h) \\
& - 2\eta_i^{\mathrm{T}}(t)P\Delta BD^{\mathrm{T}}P\eta_{i+1}(t-h) - 2\eta_i^{\mathrm{T}}(t)P\psi_i \\
\leq& \eta_i^{\mathrm{T}}(t)\left(A^{\mathrm{T}}P + PA - 2\lambda_k PDD^{\mathrm{T}}P\right)\eta_i(t) + \frac{1}{\mu}\eta_i^{\mathrm{T}}(t)PEE^{\mathrm{T}}P\eta_i(t) \\
& + \eta_i^{\mathrm{T}}(t)PDD^{\mathrm{T}}P\eta_i(t) + \eta_{i+1}^{\mathrm{T}}(t)PDD^{\mathrm{T}}P\eta_{i+1}(t) + \frac{1}{\epsilon}\lambda_k\eta_i^{\mathrm{T}}(t)PEE^{\mathrm{T}}P\eta_i(t) \\
& + \epsilon\lambda_k\eta_i^{\mathrm{T}}(t-h)PDF_2^{\mathrm{T}}F_2D^{\mathrm{T}}P\eta_i(t-h) + \frac{1}{\epsilon}\eta_i^{\mathrm{T}}(t)PEE^{\mathrm{T}}P\eta_i(t) \\
& + \epsilon\eta_{i+1}^{\mathrm{T}}(t-h)PDF_2^{\mathrm{T}}F_2D^{\mathrm{T}}P\eta_{i+1}(t-h) - 2\eta_i^{\mathrm{T}}(t)P\psi_i + \mu\eta_i^{\mathrm{T}}F_1^{\mathrm{T}}F_1\eta_i \\
=& \eta_i^{\mathrm{T}}\left[A^{\mathrm{T}}P + PA - (2\lambda_k - 1)PDD^{\mathrm{T}}P + \left(\frac{\lambda_k+1}{\epsilon} + \frac{1}{\mu}\right)PEE^{\mathrm{T}}P + \mu F_1^{\mathrm{T}}F_1\right]\eta_i \\
& + \eta_{i+1}^{\mathrm{T}}(t)PDD^{\mathrm{T}}P\eta_i(t) + \epsilon(1+\lambda_k)\eta_{i+1}^{T}(t-h)PDF_2^{\mathrm{T}}F_2D^{\mathrm{T}}P\eta_{i+1}(t-h) \\
& - 2\eta_i^{\mathrm{T}}(t)P\psi_i,
\end{aligned} \tag{3.24}$$

and, for $i = N_k$, the time derivative of W_i along the trajectory (3.21) is

$$\begin{aligned}
\dot{W}_i =& \eta_i^{\mathrm{T}}(t)P\dot{\eta}_i(t) \\
=& \eta_i^{\mathrm{T}}(t)\left(A^{\mathrm{T}}P + PA - 2\lambda_k PDD^{\mathrm{T}}P\right)\eta_i(t) + 2\eta_i^{\mathrm{T}}(t)P\Delta A\eta_i(t) \\
& - 2\lambda_k\eta_i^{\mathrm{T}}(t)P\Delta BD^{\mathrm{T}}P\eta_i(t-h) - 2\eta_i^{\mathrm{T}}(t)P\psi_i \\
\leq& \eta_i^{\mathrm{T}}(t)\left(A^{\mathrm{T}}P + PA - 2\lambda_k PDD^{\mathrm{T}}P\right)\eta_i(t) + \frac{1}{\mu}\eta_i^{\mathrm{T}}(t)PEE^{\mathrm{T}}P\eta_i(t) \\
& + \mu\eta_i^{\mathrm{T}}(t)F_1^{\mathrm{T}}F_1\eta_i(t) + \frac{1}{\epsilon}\lambda_k\eta_i^{\mathrm{T}}(t)PEE^{\mathrm{T}}P\eta_i(t) \\
& + \epsilon\lambda_k\eta_i^{\mathrm{T}}(t-h)PDF_2^{\mathrm{T}}F_2D^{\mathrm{T}}P\eta_i(t-h) - 2\eta_i^{\mathrm{T}}(t)P\psi_i \\
=& \eta_i^{\mathrm{T}}\left[A^{\mathrm{T}}P + PA - 2\lambda_k PDD^{\mathrm{T}}P + \left(\frac{\lambda_k}{\epsilon} + \frac{1}{\mu}\right)PEE^{\mathrm{T}}P + \mu F_1^{\mathrm{T}}F_1\right]\eta_i \\
& + \epsilon\lambda_k\eta_i^{\mathrm{T}}(t-h)PDF_2^{\mathrm{T}}F_2D^{\mathrm{T}}P\eta_i(t-h) - 2\eta_i^{\mathrm{T}}(t)P\psi_i,
\end{aligned} \tag{3.25}$$

where the inequality $\pm a^{\mathrm{T}}b \leq a^{\mathrm{T}}a + b^{\mathrm{T}}b$, for any vectors a and b, has been used.

Similarly, for $j = n_k/2$ in (3.22)[1], we have in pairs

$$\begin{aligned}
& \dot{W}_{i_1} + \dot{W}_{i_2} \\
& \leq \eta_{i_1}^{\mathrm{T}}\left[A^{\mathrm{T}}P + PA - 2\alpha_k PDD^{\mathrm{T}}P + \left(\frac{\alpha_k+\beta_k}{\epsilon} + \frac{1}{\mu}\right)PEE^{\mathrm{T}}P + \mu F_1^{\mathrm{T}}F_1\right]\eta_{i_1}
\end{aligned}$$

[1]We omit without ambiguity the parameter j in $i_1(j)$ and $i_2(j)$ for simplicity.

$$+\eta_{i_2}^{\mathrm{T}}\left[A^{\mathrm{T}}P+PA-2\alpha_k PDD^{\mathrm{T}}P+\left(\frac{\alpha_k+\beta_k}{\epsilon}+\frac{1}{\mu}\right)PEE^{\mathrm{T}}P+\mu F_1^{\mathrm{T}}F_1\right]\eta_{i_2}$$
$$+\epsilon(\alpha_k+\beta_k)\eta_{i_1}^{\mathrm{T}}(t-h)PDF_2^{\mathrm{T}}F_2D^{\mathrm{T}}P\eta_{i_1}(t-h)-2\eta_{i_1}(t)^{\mathrm{T}}P\psi_{i_1}$$
$$+\epsilon(\alpha_k+\beta_k)\eta_{i_2}^{\mathrm{T}}(t-h)PDF_2^{\mathrm{T}}F_2D^{\mathrm{T}}P\eta_{i_2}(t-h)-2\eta_{i_2}(t)^{\mathrm{T}}P\psi_{i_2}, \quad (3.26)$$

and, for $j=1,2,\ldots,n_k/2-1$,

$$\begin{aligned}
&\dot{W}_{i_1}+\dot{W}_{i_2}\\
&\le \eta_{i_1}^{\mathrm{T}}(t)\left[A^{\mathrm{T}}P+PA-2(\alpha_k-1)PDD^{\mathrm{T}}P+\tilde{\mu}PEE^{\mathrm{T}}P+\mu F_1^{\mathrm{T}}F_1\right]\eta_{i_1}(t)\\
&+\eta_{i_2}^{\mathrm{T}}(t)\left[A^{\mathrm{T}}P+PA-2(\alpha_k-1)PDD^{\mathrm{T}}P+\tilde{\mu}PEE^{\mathrm{T}}P+\mu F_1^{\mathrm{T}}F_1\right]\eta_{i_2}(t)\\
&+\epsilon(\alpha_k+\beta_k)\eta_{i_1}^{\mathrm{T}}(t-h)PDF_2^{\mathrm{T}}F_2D^{\mathrm{T}}P\eta_{i_1}(t-h)+\eta_{i_1+2}^{\mathrm{T}}(t)PDD^{\mathrm{T}}P\eta_{i_1+2}(t)\\
&+\epsilon(\alpha_k+\beta_k)\eta_{i_2}^{\mathrm{T}}(t-h)PDF_2^{\mathrm{T}}F_2D^{\mathrm{T}}P\eta_{i_2}(t-h)+\eta_{i_2+2}^{\mathrm{T}}(t)PDD^{\mathrm{T}}P\eta_{i_2+2}(t)\\
&+\epsilon\eta_{i_1+2}^{\mathrm{T}}(t-h)PDF_2^{\mathrm{T}}F_2D^{\mathrm{T}}P\eta_{i_1+2}(t-h)-2\eta_{i_1}(t)^{\mathrm{T}}P\psi_{i_1}\\
&+\epsilon\eta_{i_2+2}^{\mathrm{T}}(t-h)PDF_2^{\mathrm{T}}F_2D^{\mathrm{T}}P\eta_{i_2+2}(t-h)-2\eta_{i_2}(t)^{\mathrm{T}}P\psi_{i_2}. \qquad (3.27)
\end{aligned}$$

where $\tilde{\mu}=\left(\frac{\alpha_k+\beta_k+1}{\epsilon}+\frac{1}{\mu}\right)$ and the inequality $\pm a^{\mathrm{T}}b\le a^{\mathrm{T}}a+b^{\mathrm{T}}b$, for any vectors a and b, has been used again.

The above inequalities will be used in the consensus analysis. However, we note that the extra integral term $\psi_i(z)$ in the transformed system dynamic model (3.16) is expressed as a function of the state z. For the consensus analysis within the framework of Lyapunov-Krasovskii functionals, we need to establish a bound of the integral function $-2\eta_i^{\mathrm{T}}P\psi_i$ in terms of the transformed state η. The following lemma establishes a bound for the cross term $-2\eta_i^{\mathrm{T}}P\psi_i$ with respect to the transformed state η.

Lemma 3.2.1 *For the integral term $\Psi(z)=[\psi_1(z),\psi_2(z),\ldots,\psi_N(z)]^{\mathrm{T}}$ in (3.16), the summation of $-2\eta_i^{\mathrm{T}}P\psi_i$ is bounded by*

$$-\sum_{i=1}^{N}2\eta_i^{\mathrm{T}}P\psi_i\le\frac{2}{\rho}\sum_{i=2}^{N}\eta_i^{\mathrm{T}}(t)PEE^{\mathrm{T}}P\eta_i(t)$$
$$+\rho\gamma_0^2\sum_{i=2}^{N}h\int_0^h\eta_l^{\mathrm{T}}(t-\tau)PDD^{\mathrm{T}}e^{A^{\mathrm{T}}\tau}F_1^{\mathrm{T}}F_1e^{A\tau}DD^{\mathrm{T}}P\eta_l(t-\tau)\mathrm{d}\tau, \quad (3.28)$$

where $\gamma_0^2=2\|T^{-1}\|_{\mathrm{F}}^2(1+N\|r\|_2^2)\|Q\|_{\mathrm{F}}^2\|T\|_{\mathrm{F}}^2$.

Proof 9 *From the state transformations (3.13) and (3.15), we have*

$$\Psi(z)=\left(T^{-1}\otimes I_n\right)(M\otimes I_n)(I_N\otimes\Delta A)\sigma.$$

Let

$$\Phi=[\phi_1,\ldots,\phi_N]^{\mathrm{T}}=(M\otimes I_n)\bar{\sigma},$$
$$\bar{\sigma}=[\bar{\sigma}_1,\ldots,\bar{\sigma}_N]^{\mathrm{T}}=(I_N\otimes\Delta A)\sigma.$$

Recalling $M = I_N - \mathbf{1}r^{\mathrm{T}}$, *we have*

$$\phi_k = \bar{\sigma}_k - \sum_{j=1}^{N} r_j \bar{\sigma}_j = \Delta A \left(\sigma_k - \sum_{j=1}^{N} r_j \sigma_j \right),$$

$$\begin{aligned}\psi_i = (\tau_i \otimes I_n)\Phi = \sum_{k=1}^{N} \tau_{ik}\phi_k &= \sum_{k=1}^{N} \tau_{ik} \Delta A \left(\sigma_k - \sum_{j=1}^{N} r_j \sigma_j \right) \\ &= \Delta A \sum_{k=1}^{N} \tau_{ik}\sigma_k - \Delta A \sum_{k=1}^{N} \tau_{ik} \sum_{j=1}^{N} r_j \sigma_j,\end{aligned}$$

where τ_i *is the* i*th row of* T^{-1}. *It then follows that*

$$\begin{aligned}-2\eta_i^{\mathrm{T}} P \psi_i =& 2\eta_i^{\mathrm{T}} P \Delta A \sum_{k=1}^{N} \tau_{ik} \sum_{j=1}^{N} r_j \sigma_j - 2\eta_i^{\mathrm{T}} P \Delta A \sum_{k=1}^{N} \tau_{ik}\sigma_k \\ \leq& \frac{2}{\rho} \eta_i^{\mathrm{T}} P E E^{\mathrm{T}} P \eta_i + \rho \left(\sum_{k=1}^{N} \tau_{ik}\sigma_k \right)^{\mathrm{T}} F_1^{\mathrm{T}} F_1 \left(\sum_{k=1}^{N} \tau_{ik}\sigma_k \right) \\ &+ \rho \left(\sum_{k=1}^{N} \tau_{ik} \sum_{j=1}^{N} r_j \sigma_j \right)^{\mathrm{T}} F_1^{\mathrm{T}} F_1 \left(\sum_{k=1}^{N} \tau_{ik} \sum_{j=1}^{N} r_j \sigma_j \right) \\ \leq& \frac{2}{\rho} \eta_i^{\mathrm{T}} P E E^{\mathrm{T}} P \eta_i + \rho \left(\|\tau_i\|_2^2 + \|\tau_i\|_1^2 \|r\|_2^2 \right) \sum_{k=1}^{N} \|F_1 \sigma_k\|_2^2. \qquad (3.29)\end{aligned}$$

From (3.7) and (3.11), we have

$$\begin{aligned}\sigma_k =& -\int_t^{t+h} \mathrm{e}^{A(t-\tau)} BK \sum_{j=1}^{N} l_{kj} z_j(\tau - h) \mathrm{d}\tau \\ =& \int_t^{t+h} \mathrm{e}^{A(t-\tau)} BK \sum_{j=1}^{N} q_{kj} \left[z_j(\tau - h) - z_k(\tau - h) \right] \mathrm{d}\tau \\ =& \int_t^{t+h} \mathrm{e}^{A(t-\tau)} BK \sum_{j=1}^{N} q_{kj} \left[(t_j - t_k) \otimes I_n \right] \eta(\tau - h) \mathrm{d}\tau \\ =& \int_t^{t+h} \mathrm{e}^{A(t-\tau)} BK \sum_{j=1}^{N} q_{kj} \sum_{l=1}^{N} (t_{jl} - t_{kl}) \eta_l(\tau - h) \mathrm{d}\tau \\ =& \sum_{j=1}^{N} q_{kj} \sum_{l=1}^{N} (t_{jl} - t_{kl}) \delta_l, \qquad (3.30)\end{aligned}$$

where t_i is the ith row of T and

$$\delta_l = \int_t^{t+h} \mathrm{e}^{A(t-\tau)} BK\eta_l(\tau - h)\mathrm{d}\tau. \tag{3.31}$$

It then follows that

$$\begin{aligned}\sum_{k=1}^{N} \|F_1\sigma_k\|_2^2 &= \sum_{k=1}^{N} \| \sum_{j=1}^{N} q_{kj} \sum_{l=1}^{N} (t_{jl} - t_{kl}) F_1\delta_l\|_2^2 \\ &\leq \sum_{k=1}^{N} \| \sum_{j=1}^{N} q_{kj} \sum_{l=1}^{N} t_{jl} F_1\delta_l\|_2^2 + \sum_{k=1}^{N} \| \sum_{j=1}^{N} q_{kj} \sum_{l=1}^{N} t_{kl} F_1\delta_l\|_2^2 \\ &\leq 2\|Q\|_{\mathrm{F}}^2 \|T\|_{\mathrm{F}}^2 \sum_{l=1}^{N} \|F_1\delta_l\|_2^2. \end{aligned} \tag{3.32}$$

We next deal with $\|F_1\delta_l\|_2^2$ in (3.32). Then, using Lemma 2.1.2 we have

$$\begin{aligned}\|F_1\delta_l\|_2^2 &= \delta_l^{\mathrm{T}} F_1^{\mathrm{T}} F_1 \delta_l \\ &= \left(\int_t^{t+h} F_1 \mathrm{e}^{A(t-\tau)} BK\eta_l(\tau - h)\mathrm{d}\tau\right)^{\mathrm{T}} \left(\int_t^{t+h} F_1 \mathrm{e}^{A(t-\tau)} BK\eta_l(\tau - h)\mathrm{d}\tau\right) \\ &\leq h \int_t^{t+h} \eta_l^{\mathrm{T}}(\tau - h) K^{\mathrm{T}} D^{\mathrm{T}} \mathrm{e}^{A^{\mathrm{T}}(t-\tau+h)} F_1^{\mathrm{T}} F_1 \mathrm{e}^{A(t-\tau+h)} DK\eta_l(\tau - h)\mathrm{d}\tau \\ &= h \int_0^h \eta_l^{\mathrm{T}}(t - \tau) PDD^{\mathrm{T}} \mathrm{e}^{A^{\mathrm{T}}\tau} F_1^{\mathrm{T}} F_1 \mathrm{e}^{A\tau} DD^{\mathrm{T}} P\eta_l(t - \tau)\mathrm{d}\tau. \end{aligned} \tag{3.33}$$

With (3.32)-(3.33) and $\eta_1 \equiv 0$, the summation of $-2\eta_i^T P\psi_i$ can be obtained as

$$\begin{aligned} -\sum_{i=2}^{N} 2\eta_i^{\mathrm{T}} P\psi_i &\leq \frac{2}{\rho} \sum_{i=2}^{N} \eta_i^{\mathrm{T}}(t) PEE^{\mathrm{T}} P\eta_i(t) \\ &+ 2\rho \sum_{i=2}^{N} \left(\|\tau_i\|_2^2 + \|\tau_i\|_1^2 \|r\|_2^2\right) \|Q\|_{\mathrm{F}}^2 \|T\|_{\mathrm{F}}^2 \sum_{l=2}^{N} \|F_1\delta_l\|_2^2 \\ \leq& \frac{2}{\rho} \sum_{i=2}^{N} \eta_i^{\mathrm{T}}(t) PEE^{\mathrm{T}} P\eta_i(t) \\ &+ \rho\gamma_0^2 \sum_{i=2}^{N} h \int_0^h \eta_l^{\mathrm{T}}(t - \tau) PDD^{\mathrm{T}} \mathrm{e}^{A^{\mathrm{T}}\tau} F_1^{\mathrm{T}} F_1 \mathrm{e}^{A\tau} DD^{\mathrm{T}} P\eta_l(t - \tau)\mathrm{d}\tau, \end{aligned}$$

where $2\sum_{i=2}^{N}(\|\tau_i\|_2^2 + \|\tau_i\|_1^2\|r\|_2^2)\|Q\|_{\mathrm{F}}^2\|T\|_{\mathrm{F}}^2 \leq 2\|T^{-1}\|_{\mathrm{F}}^2(1 + N\|r\|_2^2)\|Q\|_{\mathrm{F}}^2\|T\|_{\mathrm{F}}^2$ has been inserted in the last inequality with $\sum_{i=1}^{N} \|\tau_i\|_2^2 = \|T^{-1}\|_{\mathrm{F}}^2$ being used.

With the bound derived in Lemma 3.2.1, sufficient conditions can be identified respectively for the cases of the Laplacian matrix with distinct eigenvalues and multiple eigenvalues to guarantee the consensus. The following theorem summarises the results.

Theorem 7 *Consider multi-agent systems (3.1) with Assumptions 3.1.1 and 3.1.2. The consensus control problem of system (3.1) can be solved by the control design (3.7) with the control gain $K = D^{\mathrm{T}}P$, if there exist matrices $X = P^{-1} > 0, Y > 0$ and scalars $\mu > 0, \epsilon > 0, \rho > 0$, such that*

$$\begin{bmatrix} Y & DF_2^{\mathrm{T}} \\ F_2D^{\mathrm{T}} & \frac{1}{\epsilon}I \end{bmatrix} > 0, \tag{3.34}$$

$$\begin{bmatrix} U & XF_1^{\mathrm{T}} & DD^T \\ F_1X & -\frac{1}{\mu}I & 0 \\ DD^{\mathrm{T}} & 0 & -\frac{1}{\rho\gamma_0^2}W \end{bmatrix} < 0, \tag{3.35}$$

where U is specified in one of the following two cases:

(i) If the eigenvalues of the Laplacian matrix $\mathcal{L}$ are distinct,

$$U = XA^{\mathrm{T}} + AX - 2\underline{\alpha}DD^{\mathrm{T}} + \left(\frac{1}{\mu} + \frac{\bar{\alpha}+\bar{\beta}}{\epsilon} + \frac{2}{\rho}\right)EE^{\mathrm{T}} + (\bar{\alpha}+\bar{\beta})Y. \tag{3.36}$$

(ii) If the Laplacian matrix $\mathcal{L}$ has multiple eigenvalues,

$$U = XA^{\mathrm{T}} + AX - 2(\underline{\alpha}-1)DD^{\mathrm{T}} + \left(\frac{1}{\mu} + \frac{\bar{\alpha}+\bar{\beta}+1}{\epsilon} + \frac{2}{\rho}\right)EE^{\mathrm{T}} + (\bar{\alpha}+\bar{\beta}+1)Y, \tag{3.37}$$

and

$$\gamma_0^2 = 2\|T^{-1}\|_{\mathrm{F}}^2(1+N\|r\|_2^2)\|Q\|_{\mathrm{F}}^2\|T\|_{\mathrm{F}}^2, \tag{3.38}$$

$$W^{-1} \geq h\int_0^h \mathrm{e}^{A^{\mathrm{T}}s}F_1^{\mathrm{T}}F_1\mathrm{e}^{As}\mathrm{d}s, \tag{3.39}$$

$$\bar{\alpha} = \max\{\lambda_2,\ldots,\lambda_{n_\lambda},\alpha_1,\ldots,\alpha_{n_\nu}\}, \tag{3.40}$$

$$\bar{\beta} = \max\{\beta_1,\ldots,\beta_{n_\nu}\}, \tag{3.41}$$

$$\underline{\alpha} = \min\{\lambda_2,\ldots,\lambda_{n_\lambda},\alpha_1,\ldots,\alpha_{n_\nu}\}. \tag{3.42}$$

Proof 10 *For all the state variables associate with the Jordan blocks of real eigenvalues, we consider the following summation of* (3.23)*:*

$$V_k = \sum_{j=1}^{n_k} W_{j+N_{k-1}}, \tag{3.43}$$

and from (3.24)–(3.25) *we then obtain*

$$\begin{aligned}\dot{V}_k \leq & \sum_{j=1}^{n_k}\eta_{j+N_{k-1}}^{\mathrm{T}}(t)\left[A_{11} + \left(\frac{\lambda_k+1}{\epsilon} + \frac{1}{\mu}\right)PEE^{\mathrm{T}}P + \mu F_1^{\mathrm{T}}F_1\right]\eta_{j+N_{k-1}}(t) \\ & - \eta_{1+N_{k-1}}^{\mathrm{T}}(t)PDD^{\mathrm{T}}P\eta_{1+N_{k-1}}(t) - \eta_{N_k}^{\mathrm{T}}(t)PDD^{\mathrm{T}}P\eta_{N_k}(t)\end{aligned}$$

$$+\sum_{j=1}^{n_k}\epsilon(1+\lambda_k)\eta_{j+N_{k-1}}^{\mathrm{T}}(t-h)PDF_2^{\mathrm{T}}F_2D^{\mathrm{T}}P\eta_{j+N_{k-1}}(t-h)$$

$$-\epsilon\eta_{1+N_{k-1}}^{\mathrm{T}}(t-h)PDF_2^{\mathrm{T}}F_2D^{\mathrm{T}}P\eta_{1+N_{k-1}}(t-h)-2\sum_{j=1}^{n_k}\eta_{j+N_{k-1}}^{\mathrm{T}}(t)P\psi_i$$

$$-\frac{1}{\epsilon}\eta_{N_k}^{\mathrm{T}}(t)PEE^{\mathrm{T}}P\eta_{N_k}(t), \tag{3.44}$$

where $A_{11} = A^{\mathrm{T}}P + PA - 2(\lambda_k - 1)PDD^{\mathrm{T}}P$. *For all the state variables corresponding to the conjugate eigenvalues in the Jordan blocks, we consider the following summation of* (3.23) *in pairs:*

$$V_k = \sum_{j=1}^{n_k/2}\left[W_{i_1(j)} + W_{i_2(j)}\right]. \tag{3.45}$$

By (3.26)–(3.27) *and after simple mechanical calculation, we have*

$$\begin{aligned}\dot{V}_k &= \sum_{j=1}^{n_k/2}\left[\dot{W}_{i_1(j)} + \dot{W}_{i_2(j)}\right]\\
&\leq \sum_{j=1}^{n_k/2}\eta_{2j-1+N_{k-1}}^{\mathrm{T}}(t)\left[A_{12}+\mu_{12}PEE^{\mathrm{T}}P+\mu F_1^{\mathrm{T}}F_1\right]\eta_{2j-1+N_{k-1}}(t)\\
&\quad+\sum_{j=1}^{n_k/2}\eta_{2j+N_{k-1}}^{\mathrm{T}}(t)\left[A_{12}+\mu_{12}PEE^{\mathrm{T}}P+\mu F_1^{\mathrm{T}}F_1\right]\eta_{2j+N_{k-1}}(t)\\
&\quad-\sum_{j=1}^{n_k/2}2\eta_{2j-1+N_{k-1}}^{\mathrm{T}}(t)P\psi_{2j-1+N_{k-1}}-\sum_{j=1}^{n_k/2}2\eta_{2j+N_{k-1}}^{\mathrm{T}}(t)P\psi_{2j+N_{k-1}}\\
&\quad-\eta_{1+N_{k-1}}^{\mathrm{T}}(t)PDD^{\mathrm{T}}P\eta_{1+N_{k-1}}(t)-\eta_{2+N_{k-1}}^{\mathrm{T}}(t)PDD^{\mathrm{T}}P\eta_{2+N_{k-1}}(t)\\
&\quad-\eta_{N_k-1}^{\mathrm{T}}(t)PDD^{\mathrm{T}}P\eta_{N_k-1}(t)-\eta_{N_k}^{\mathrm{T}}(t)PDD^{\mathrm{T}}P\eta_{N_k}(t)\\
&\quad-\frac{1}{\epsilon}\eta_{N_k-1}^{\mathrm{T}}(t)PEE^{\mathrm{T}}P\eta_{N_k-1}(t)-\frac{1}{\epsilon}\eta_{N_k}^{\mathrm{T}}(t)PEE^{\mathrm{T}}P\eta_{N_k}(t)\\
&\quad+\sum_{j=1}^{n_k/2}\epsilon(\alpha_k+\beta_k+1)\eta_{2j-1+N_{k-1}}^{\mathrm{T}}(t-h)PDF_2^{\mathrm{T}}F_2D^{\mathrm{T}}P\eta_{2j-1+N_{k-1}}(t-h)\\
&\quad+\sum_{j=1}^{n_k/2}\epsilon(\alpha_k+\beta_k+1)\eta_{2j+N_{k-1}}^{\mathrm{T}}(t-h)PDF_2^{\mathrm{T}}F_2D^{\mathrm{T}}P\eta_{2j+N_{k-1}}(t-h)\\
&\quad-\epsilon\eta_{N_k-1}^{\mathrm{T}}(t-h)PDF_2^{\mathrm{T}}F_2D^{\mathrm{T}}P\eta_{N_k-1}(t-h)\\
&\quad-\epsilon\eta_{N_k}^{\mathrm{T}}(t-h)PDF_2^{\mathrm{T}}F_2D^{\mathrm{T}}P\eta_{N_k}(t-h),\end{aligned} \tag{3.46}$$

where $A_{12} = A^{\mathrm{T}}P + PA - 2(\lambda_k - 1)PDD^{\mathrm{T}}P$, $\mu_{12} = \left(\dfrac{\alpha_k+\beta_k+1}{\epsilon}+\dfrac{1}{\mu}\right)$.

With (3.43) *and* (3.45), *we consider the following Lyapunov function*

$$\begin{aligned}\mathcal{V}_1 &= \sum_{k=2}^{q} V_k = \sum_{k=2}^{p} V_k + \sum_{k=p+1}^{q} V_k \\ &= \sum_{k=2}^{p}\sum_{j=1}^{n_k} W_{j+N_{k-1}} + \sum_{k=p+1}^{q}\sum_{j=1}^{n_k/2}\left[W_{i_1(j)} + W_{i_2(j)}\right],\end{aligned} \tag{3.47}$$

where N_k *is defined in* (3.19) *with* $N_1 = n_1 = 1$ *and* $2 \le k \le q$, $i_1(j)$ *and* $i_2(j)$ *are defined in* (3.22). *By* (3.44) *and* (3.46), *we can compute the time derivative of* (3.47) *as*

$$\begin{aligned}\dot{\mathcal{V}}_1 \le& \sum_{k=2}^{p}\sum_{j=1}^{n_k} \eta_{j+N_{k-1}}^{\mathrm{T}}(t)\left[A_{12} + \left(\frac{\lambda_k+1}{\epsilon} + \frac{1}{\mu}\right)P_1^* + \mu F_1^{\mathrm{T}} F_1\right]\eta_{j+N_{k-1}}(t) \\ &+ \sum_{k=p+1}^{q}\sum_{j=1}^{n_k/2} \eta_{2j-1+N_{k-1}}^{\mathrm{T}}(t)\left[A_{12} + \epsilon_{12} P_1^* + \mu F_1^{\mathrm{T}} F_1\right]\eta_{2j-1+N_{k-1}}(t) \\ &+ \sum_{k=p+1}^{q}\sum_{j=1}^{n_k/2} \eta_{2j+N_{k-1}}^{\mathrm{T}}(t)\left[A_{12} + \epsilon_{12} P_1^* + \mu F_1^{\mathrm{T}} F_1\right]\eta_{2j+N_{k-1}}(t) \\ &- \sum_{k=2}^{p}\underbrace{\left[\eta_{1+N_{k-1}}^{\mathrm{T}}(t)P_2^*\eta_{1+N_{k-1}}(t) + \eta_{N_k}^{\mathrm{T}}(t)P_2^*\eta_{N_k}(t)\right]}_{\ge 0} \\ &- \sum_{k=p+1}^{q}\underbrace{\left[\eta_{1+N_{k-1}}^{\mathrm{T}}(t)P_2^*\eta_{1+N_{k-1}}(t) + \eta_{2+N_{k-1}}^{\mathrm{T}}(t)P_2^*\eta_{2+N_{k-1}}(t)\right]}_{\ge 0} \\ &- \sum_{k=p+1}^{q}\underbrace{\left[\eta_{N_k-1}^{\mathrm{T}}(t)P_2^*\eta_{N_k-1}(t) + \eta_{N_k}^{\mathrm{T}}(t)P_2^*\eta_{N_k}(t)\right]}_{\ge 0} \\ &- \sum_{k=p+1}^{q}\frac{1}{\epsilon}\underbrace{\left[\eta_{N_k-1}^{\mathrm{T}}(t)P_1^*\eta_{N_k-1}(t) + \eta_{N_k}^{\mathrm{T}}(t)P_1^*\eta_{N_k}(t)\right]}_{\ge 0} \\ &+ \sum_{k=2}^{p}\sum_{j=1}^{n_k} \epsilon(1+\lambda_k)\eta_{j+N_{k-1}}^{\mathrm{T}}(t-h)PDF_2^{\mathrm{T}}F_2D^{\mathrm{T}}P\eta_{j+N_{k-1}}(t-h) \\ &+ \sum_{k=p+1}^{q}\sum_{j=1}^{n_k/2} \epsilon(\alpha_k+\beta_k+1)\eta_{2j-1+N_{k-1}}^{\mathrm{T}}(t-h)P_3^*\eta_{2j-1+N_{k-1}}(t-h)\end{aligned}$$

$$
\begin{aligned}
&+\sum_{k=p+1}^{q}\sum_{j=1}^{n_k/2}\epsilon(\alpha_k+\beta_k+1)\eta_{2j+N_{k-1}}^{\mathrm{T}}(t-h)P_3^*\eta_{2j+N_{k-1}}(t-h)\\
&-\sum_{k=2}^{p}\epsilon\underbrace{\eta_{1+N_{k-1}}^{\mathrm{T}}(t-h)P_3^*\eta_{1+N_{k-1}}(t-h)}_{\geq 0}\\
&-\sum_{k=p+1}^{q}\epsilon\underbrace{\left[\eta_{N_k-1}^{\mathrm{T}}(t-h)P_3^*\eta_{N_k-1}(t-h)+\eta_{N_k}^{\mathrm{T}}(t-h)P_3^*\eta_{N_k}(t-h)\right]}_{\geq 0}\\
&-\sum_{k=2}^{p}\frac{1}{\epsilon}\underbrace{\eta_{N_k}^{\mathrm{T}}(t)P_1^*\eta_{N_k}(t)}_{\geq 0}-2\sum_{i=2}^{N}\eta_i^{\mathrm{T}}(t)P\psi_i(z),
\end{aligned}
\tag{3.48}
$$

where $\epsilon_{12} = \dfrac{\alpha_k+\beta_k+1}{\epsilon} + \dfrac{1}{\mu}, P_1^* = PEE^{\mathrm{T}}P, P_2^* = PDD^{\mathrm{T}}P, P_3^* = PDF_2^{\mathrm{T}}F_2D^{\mathrm{T}}P, A_{12}$ *is defined in 3.46. Substituting (3.28) into (3.48) and using the definitions* (3.40)–(3.42), *we have the following two cases:*

(i) when the Laplacian matrix $\mathcal{L}$ *has distinct eigenvalues, i.e.,* $n_k = 1$ *for all* $k \in \{2, 3, \ldots, q\}$,

$$
\begin{aligned}
\dot{\mathcal{V}}_1 \leq &\sum_{i=2}^{N}\eta_i^{\mathrm{T}}(t)\left[A_{13}+\left(\frac{1}{\mu}+\frac{\bar{\alpha}+\bar{\beta}}{\epsilon}+\frac{2}{\rho}\right)P_1^*+\mu F_1^{\mathrm{T}}F_1\right]\eta_i(t)\\
&+\sum_{i=2}^{N}\epsilon(\bar{\alpha}+\bar{\beta})\eta_i^{\mathrm{T}}(t-h)PDF_2^{\mathrm{T}}F_2D^{\mathrm{T}}P\eta_i(t-h)\\
&+\rho\gamma_0^2\sum_{i=2}^{N}h\int_0^h\eta_i^{\mathrm{T}}(t-\tau)PDD^{\mathrm{T}}e^{A^{\mathrm{T}}\tau}F_1^{\mathrm{T}}F_1e^{A\tau}DD^{\mathrm{T}}P\eta_i(t-\tau)\mathrm{d}\tau,
\end{aligned}
\tag{3.49}
$$

where $A_{13} = A^{\mathrm{T}}P + PA - 2\underline{\alpha}PDD^{\mathrm{T}}P$.

(ii) when the Laplacian matrix $\mathcal{L}$ *has multiple eigenvalues, i.e.,* $n_k > 1$ *for any* $k \in \{2, 3, \ldots, q\}$,

$$
\begin{aligned}
\dot{\mathcal{V}}_1 \leq &\sum_{i=2}^{N}\eta_i^{\mathrm{T}}(t)\left[A_{13}+\left(\frac{1}{\mu}+\frac{\bar{\alpha}+\bar{\beta}+1}{\epsilon}+\frac{2}{\rho}\right)PEE^{\mathrm{T}}P+\mu F_1^{\mathrm{T}}F_1\right]\eta_i(t)\\
&+\sum_{i=2}^{N}\epsilon(\bar{\alpha}+\bar{\beta}+1)\eta_i^{\mathrm{T}}(t-h)PDF_2^{\mathrm{T}}F_2D^{\mathrm{T}}P\eta_i(t-h)\\
&+\rho\gamma_0^2\sum_{i=2}^{N}h\int_0^h\eta_i^{\mathrm{T}}(t-\tau)PDD^{\mathrm{T}}e^{A^{\mathrm{T}}\tau}F_1^{\mathrm{T}}F_1e^{A\tau}DD^{\mathrm{T}}P\eta_i(t-\tau)\mathrm{d}\tau.
\end{aligned}
\tag{3.50}
$$

For the delayed term shown in (3.49) and (3.50), we consider the following Krasovskii functionals for both cases, respectively,

$$\text{(i)}\quad \mathcal{V}_2 = \sum_{i=2}^{N}(\bar{\alpha}+\bar{\beta})\int_{t-h}^{t}\eta_i^{\mathrm{T}}(\tau)R\eta_i(\tau)\mathrm{d}\tau,$$
$$\text{(ii)}\quad \mathcal{V}_2 = \sum_{i=2}^{N}(\bar{\alpha}+\bar{\beta}+1)\int_{t-h}^{t}\eta_i^{\mathrm{T}}(\tau)R\eta_i(\tau)\mathrm{d}\tau,$$

where

$$R - \epsilon PDF_2^{\mathrm{T}}F_2D^{\mathrm{T}}P > 0. \tag{3.51}$$

A direct calculation gives, respectively, that

$$\text{(i)}\quad \dot{\mathcal{V}}_2 = \sum_{i=2}^{N}(\bar{\alpha}+\bar{\beta})[\eta_i^{\mathrm{T}}(t)R\eta_i(t) - \eta_i^{\mathrm{T}}(t-h)R\eta_i(t-h)], \tag{3.52}$$
$$\text{(ii)}\quad \dot{\mathcal{V}}_2 = \sum_{i=2}^{N}(\bar{\alpha}+\bar{\beta}+1)[\eta_i^{\mathrm{T}}(t)R\eta_i(t) - \eta_i^{\mathrm{T}}(t-h)R\eta_i(t-h)]. \tag{3.53}$$

For the integral term shown in (3.49) or (3.50), we consider the following Krasovskii functional

$$\mathcal{V}_3 = \rho h\gamma_0^2\sum_{i=2}^{N}\int_0^h\int_{t-s}^{t}\eta_i^{\mathrm{T}}(\tau)PDD^{\mathrm{T}}\mathrm{e}^{A^{\mathrm{T}}s}F_1^{\mathrm{T}}F_1\mathrm{e}^{As}DD^{\mathrm{T}}P\eta_i(\tau)\mathrm{d}\tau\mathrm{d}s.$$

A direct calculation gives that

$$\begin{aligned}\dot{\mathcal{V}}_3 = {} & \rho h\gamma_0^2\sum_{i=2}^{N}\int_0^h\eta_i^{\mathrm{T}}(t)PDD^{\mathrm{T}}\mathrm{e}^{A^{\mathrm{T}}s}F_1^{\mathrm{T}}F_1\mathrm{e}^{As}DD^{\mathrm{T}}P\eta_i(t)\mathrm{d}s \\ & - \rho h\gamma_0^2\sum_{i=2}^{N}\int_0^h\eta_i^{\mathrm{T}}(t-s)PDD^{\mathrm{T}}\mathrm{e}^{A^{\mathrm{T}}s}F_1^{\mathrm{T}}F_1\mathrm{e}^{As}DD^{\mathrm{T}}P\eta_i(t-s)\mathrm{d}s \\ \le {} & \rho\gamma_0^2\sum_{i=2}^{N}\eta_i^{\mathrm{T}}(t)PDD^{\mathrm{T}}W^{-1}DD^{\mathrm{T}}P\eta_i(t) \\ & - \rho h\gamma_0^2\sum_{i=2}^{N}\int_0^h\eta_i^{\mathrm{T}}(t-s)PDD^{\mathrm{T}}\mathrm{e}^{A^{\mathrm{T}}s}F_1^{\mathrm{T}}F_1\mathrm{e}^{As}DD^{\mathrm{T}}P\eta_i(t-s)\mathrm{d}s,\end{aligned} \tag{3.54}$$

where

$$W^{-1} \ge h\int_0^h \mathrm{e}^{A^{\mathrm{T}}s}F_1^{\mathrm{T}}F_1\mathrm{e}^{As}\mathrm{d}s. \tag{3.55}$$

Let

$$\mathcal{V}_0 = \mathcal{V}_1 + \mathcal{V}_2 + \mathcal{V}_3.$$

From (3.49), (3.52) and (3.54), we obtain

$$\dot{\mathcal{V}}_0 = \dot{\mathcal{V}}_1 + \dot{\mathcal{V}}_2 + \dot{\mathcal{V}}_3 \leq \sum_{i=2}^{N} \eta_i^{\mathrm{T}}(t) H \eta_i(t), \tag{3.56}$$

where

$$\begin{aligned} (i) \quad H \triangleq\, & A^{\mathrm{T}}P + PA - 2\underline{\alpha}PDD^{\mathrm{T}}P + \mu F_1^{\mathrm{T}}F_1 + (\bar{\alpha} + \bar{\beta})R \\ & + \left(\frac{1}{\mu} + \frac{\bar{\alpha} + \bar{\beta}}{\epsilon} + \frac{2}{\rho}\right) PEE^{\mathrm{T}}P + \rho\gamma_0^2 PDD^{\mathrm{T}}W^{-1}DD^{\mathrm{T}}P, \end{aligned} \tag{3.57}$$

$$\begin{aligned} \text{or } (ii) \quad H \triangleq\, & A^{\mathrm{T}}P + PA - 2(\underline{\alpha} - 1)PDD^{\mathrm{T}}P + \mu F_1^{\mathrm{T}}F_1 + (\bar{\alpha} + \bar{\beta} + 1)R \\ & + \left(\frac{1}{\mu} + \frac{\bar{\alpha} + \bar{\beta} + 1}{\epsilon} + \frac{2}{\rho}\right) PEE^{\mathrm{T}}P + \rho\gamma_0^2 PDD^{\mathrm{T}}W^{-1}DD^{\mathrm{T}}P. \end{aligned} \tag{3.58}$$

From the analysis above, the control (3.7) stabilises $\eta(t)$ if the conditions (3.51), (3.55) and $H < 0$ in (3.56) are satisfied. Indeed, it is easy to see that the conditions (3.51) and (3.55) are equivalent, respectively, to the conditions specified in (3.34) and (3.39) with $Y = P^{-1}RP^{-1}$. From (3.56), it can be shown that $H < 0$ is equivalent to

$$\begin{aligned} (i) \quad & P^{-1}A^{T} + AP^{-1} - 2\underline{\alpha}DD^{\mathrm{T}} + \left(\frac{1}{\mu} + \frac{\bar{\alpha} + \bar{\beta}}{\epsilon} + \frac{2}{\rho}\right) EE^{\mathrm{T}} \\ & + \mu P^{-1}F_1^{\mathrm{T}}F_1P^{-1} + (\bar{\alpha} + \bar{\beta})P^{-1}RP^{-1} + \rho\gamma_0^2 DD^{\mathrm{T}}W^{-1}DD^{\mathrm{T}} < 0, \end{aligned} \tag{3.59}$$

$$\begin{aligned} \text{or } (ii) \quad & P^{-1}A^{\mathrm{T}} + AP^{-1} - 2(\underline{\alpha} - 1)DD^{\mathrm{T}} + \left(\frac{1}{\mu} + \frac{\bar{\alpha} + \bar{\beta} + 1}{\epsilon} + \frac{2}{\rho}\right) EE^{T} \\ & + \mu P^{-1}F_1^{\mathrm{T}}F_1P^{-1} + (\bar{\alpha} + \bar{\beta} + 1)P^{-1}RP^{-1} + \rho\gamma_0^2 DD^{\mathrm{T}}W^{-1}DD^{\mathrm{T}} < 0, \end{aligned} \tag{3.60}$$

which is further equivalent to (3.35) with $X = P^{-1}$. Hence, we conclude that $\eta(t)$ converges to zero asymptotically. This completes the proof.

Remark 3.2.4 *The conditions shown in (3.34) to (3.35) can be checked by standard LMI routines for a set of fixed values R and W^{-1}. The iterative methods developed in [188] for single linear system may also be applied here.*

Remark 3.2.5 *Note from (3.36) and (3.37) that a more stringent condition is required for the case of the Laplacian matrix with multiple eigenvalues than the case with only distinct eigenvalues.*

Remark 3.2.6 *It can be seen from (3.39) that the matrix W^{-1} explicitly depends on the delay h, which implies that large input delays will lead to a difficulty in finding a feasible solution satisfying the conditions (3.34) and (3.35) simultaneously. Even if such a feasible solution P exists, a larger input delay results in a smaller P and therefore a smaller control gain K, which further implies a more sluggish consensus response.*

3.3 A Numerical Example

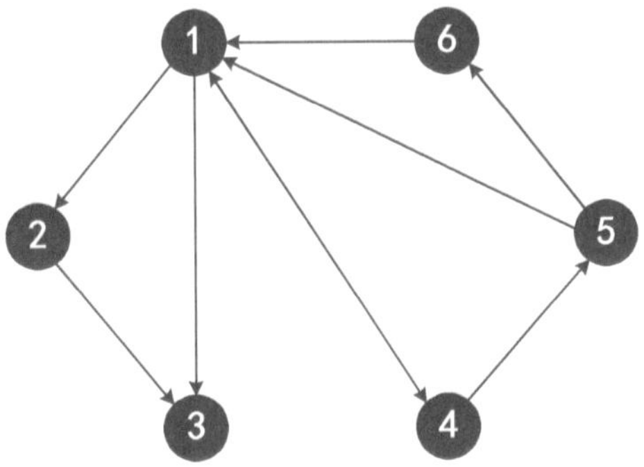

FIGURE 3.1: Network connection topology.

In this section, the scenario under consideration is a connection of six subsystems (i.e., $N = 6$) in the network as shown in Figure 3.1. The dynamics of each subsystem are described by (3.1) with

$$A = \begin{bmatrix} -1 & 1 \\ 0 & 0 \end{bmatrix}, \quad B = \begin{bmatrix} 0 \\ 1 \end{bmatrix}, \quad \Sigma(t) = \begin{bmatrix} \sin(t) & 0 \\ 0 & \sin(2t) \end{bmatrix},$$
$$E = \begin{bmatrix} 2 & 0 \\ 0 & 2 \end{bmatrix}, \quad F_1 = \begin{bmatrix} 0.1 & 0 \\ 0 & 0.1 \end{bmatrix}, \ F_2 = \begin{bmatrix} 0.1 \\ 0.1 \end{bmatrix}.$$

The Laplacian matrix associated with the graph in Figure 3.1 is

$$\mathcal{L} = \begin{bmatrix} 3 & 0 & 0 & -1 & -1 & -1 \\ -1 & 1 & 0 & 0 & 0 & 0 \\ 1 & -1 & 2 & 0 & 0 & 0 \\ -1 & 0 & 0 & 1 & 0 & 0 \\ 0 & 0 & 0 & -1 & 1 & 0 \\ 0 & 0 & 0 & 0 & -1 & 1 \end{bmatrix},$$

where the eigenvalues of $\mathcal{L}$ are $[0, 1, 2, 3.3247, 1.3376 \pm j0.5623]$, which implies

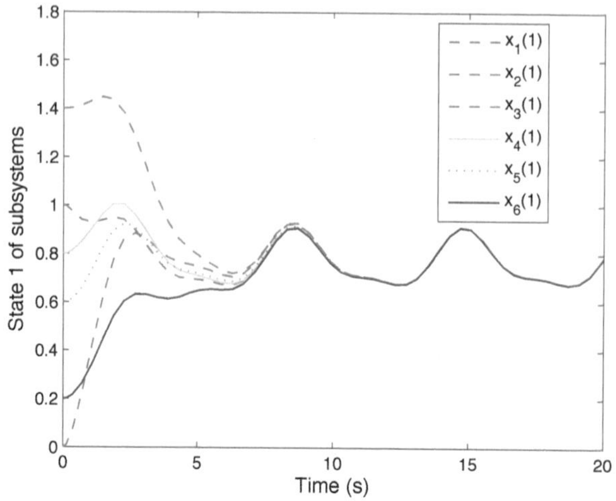

FIGURE 3.2: The state 1 of subsystems with $h = 0.03$.

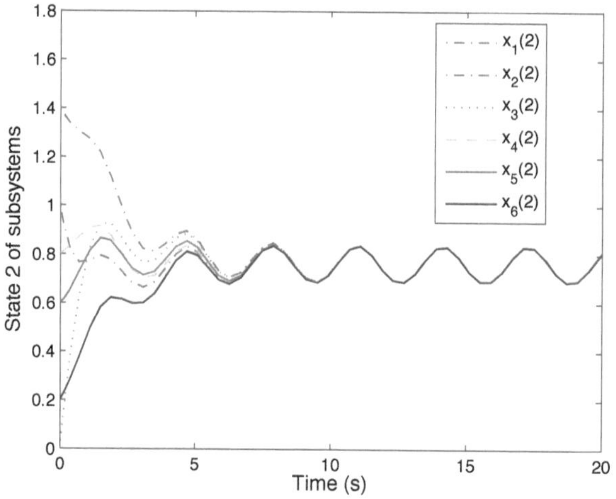

FIGURE 3.3: The state 2 of subsystems with $h = 0.03$.

the case (i) in Theorem 7 is satisfied. Then, it can be straightforward to calculate the Jordan canonical form of $\mathcal{L}$ as

$$J = \begin{bmatrix} 0 & 0 & 0 & 0 & 0 & 0 \\ 0 & 2 & 0 & 0 & 0 & 0 \\ 0 & 0 & 1 & 0 & 0 & 0 \\ 0 & 0 & 0 & 3.3247 & 0 & 0 \\ 0 & 0 & 0 & 0 & 1.3376 & 0.5623 \\ 0 & 0 & 0 & 0 & -0.5623 & 1.3376 \end{bmatrix}$$

with the matrices

$$T = \begin{bmatrix} 1 & 0 & 0 & -12.5635 & 0.2818 - j0.0145 & 0.2818 - j0.0145 \\ 1 & 0 & 1 & 5.4043 & -0.2022 + j0.3797 & -0.2022 - j0.3797 \\ 1 & 1 & 1 & 5.4043 & -0.2022 + j0.3797 & -0.2022 - j0.3797 \\ 1 & 0 & 0 & 5.4043 & -0.2022 + j0.3797 & -0.2022 - j0.3797 \\ 1 & 0 & 0 & -2.3247 & -0.3376 - j0.5623 & -0.3376 + j0.5623 \\ 1 & 0 & 0 & 1 & 1 & 1 \end{bmatrix}.$$

Thus, we have $r = [0.1429, 0, 0, 0.4286, 0.2857, 0.1429]^{\mathrm{T}}$, $\underline{\alpha} = 1$, $\bar{\alpha} = 3.3247$ and $\bar{\beta} = 0.5623$.

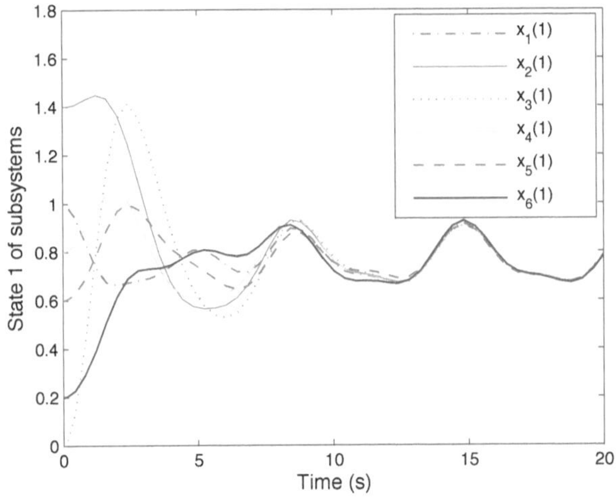

FIGURE 3.4: The state 1 of subsystems with $h = 0.3$.

The input delay of the system is $h = 0.03$ s. The positive definite matrix P can be computed with $\mu = 1$, $\epsilon = 1$ and $\rho = 0.1$, as

$$X^{-1} = P = \begin{bmatrix} 0.0002 & 0.0002 \\ 0.0002 & 0.5174 \end{bmatrix},$$

to satisfy the conditions of Theorem 7. Consequently, the control gain is obtained as

$$K = D^{\mathrm{T}} P = \begin{bmatrix} 0.0002 & 0.5173 \end{bmatrix}.$$

Simulation study has been carried out with the results shown in Figures 3.2 and 3.3 for the states of each subsystem. Clearly the conditions specified in Theorem 1 are sufficient for the control gain to achieve consensus control. Without any re-tuning the control gain, the consensus control is still achieved for the multi-agent systems with a much larger delay $h = 0.3$ s, as shown in Figures 3.4 and 3.5, which imply that the conditions could be conservative in the control gain design for a given input delay.

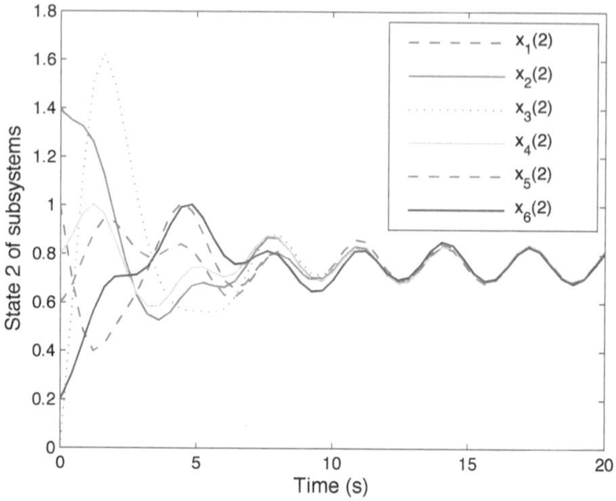

FIGURE 3.5: The state 2 of subsystems with $h = 0.3$.

3.4 Conclusions

In this chapter, we have solved the consensus problem of multi-agent systems in the presence of parametric uncertainty and input delay by exploiting the reduction method for delay together with consensus control design based on real Jordan form of the Laplacian matrix. Further analysis has been developed to tackle the influence of the extra integral term under transformations due to the model uncertainty. Sufficient conditions are derived for the closed-loop system to achieve global consensus using Lyapunov-Krasovskii method in the time domain. The significance of this research is to provide a feasible method to deal with the robust consensus control for input-delayed uncertain multi-agent systems.

3.5 Notes

This chapter develops consensus protocols for uncertain multi-agent systems with input delay and directed topology. The main materials of this chapter are based on [209] and [210]. It is worth mentioning that linear multi-agent systems are considered in this chapter. The results in this chapter can be extended to Lipschitz nonlinear systems by following the procedures shown in [211].

Chapter 4

H_∞ Consensus Control of Linear Multi-Agent Systems with Input Delay

In this chapter, the consensus problem for linear multi-agent systems with general directed graph in the presence of constant input delay and external disturbances is addressed. To deal with input delay, a truncated prediction of the agent state over the delay period is approximated by the finite dimensional term of the classical state predictor. The truncated predictor feedback method is used for the consensus protocol design. By exploring certain features of the Laplacian matrix, the H_∞ consensus analysis is put in the framework of Lyapunov analysis. The integral terms that remain in the transformed systems are carefully analyzed by using Krasovskii functional. Sufficient conditions are derived for the multi-agent systems to guarantee the H_∞ consensus in the time domain. The feedback gain is then designed by solving these conditions with an iterative LMI procedure. A simulation study is carried out to validate the proposed control design.

4.1 Problem Formulation

Consider a group of N agents, each represented by a linear dynamic subject to input delay and external disturbance,

$$\dot{x}_i(t) = Ax_i(t) + Bu_i(t-h) + D\omega_i, \tag{4.1}$$

where for agent i, $i = 1, 2, \ldots, N$, $x_i \in \mathbb{R}^n$ is the state vector, $u_i \in \mathbb{R}^{m\times n}$ is the control input vector, $A \in \mathbb{R}^{n\times n}$, $B \in \mathbb{R}^{n\times m}$ and $D \in \mathbb{R}^{n\times m}$ are constant matrices with (A, B) being controllable, $h > 0$ is input delay, and $\omega_i \in \mathcal{L}_2^m\,[0,\infty)$ is the external disturbance.

The communications among the agents are described by a directed graph with the following assumption.

Assumption 4.1.1 *Zero is a simple eigenvalue of the Laplacian matrix $\mathcal{L}$.*

Remark 4.1.1 *This assumption implies that the directed graph contains a* *spanning tree, which is essential for consensus control. If zero is not a simple eigenvalue of $\mathcal{L}$, the agents cannot reach consensus asymptotically as there exist at least two separate subgroups or at least two agents in the group who do not receive any information [125].*

Remark 4.1.2 *The left eigenvector r of the Laplacian matrix $\mathcal{L}$ with the zero eigenvalue is crucial for the consensus design with directed graph. Feasible methods are given in [89, 125] to calculate this vector. In addition, the elements of r could be zero. This suggests that the methods developed in [72] and [75] may not be suitable for the H_∞ consensus analysis here.*

The objective of this chapter is to design a control algorithm for each agent such that the multi-agent systems (4.1) achieve consensus and meanwhile maintain a desirable disturbance rejection performance. In view of this, we introduce a state transformation

$$\xi_i(t) = x_i(t) - \sum_{j=1}^{N} r_j x_j(t), \tag{4.2}$$

where $i = 1, 2, \ldots, N$, $\xi_i \in \mathbb{R}^n$, r_j denotes the jth element of r. Based on the new variable ξ_i, we define the performance variable as $e_i(t) = C\xi_i(t)$, where $e_i \in \mathbb{R}^m$, $C \in \mathbb{R}^{m\times n}$ is a constant matrix. Let $e(t) = \left[e_1^T, e_2^T, \cdots, e_N^T\right]^T$ and $\omega = \left[\omega_1^T, \omega_2^T, \cdots, \omega_N^T\right]^T$. The H_∞ consensus control problem can be defined as below.

Definition 12 *Given a positive scalar γ, the H_∞ consensus is achieved if the two requirements listed below are satisfied:*

1. The multi-agent systems (4.1) with $\omega_i \equiv 0$ can reach consensus. That is, under these control algorithms, the following hold for all initial conditions,

$$\lim_{t\to\infty} (x_i(t) - x_j(t)) = 0, \quad \forall i \neq j.$$

2. Under the zero-initial condition, the performance variable $e(t)$ satisfies

$$J = \int_0^\infty \left[e^T(t)e(t) - \gamma^2\omega^T(t)\omega(t)\right] \mathrm{d}\tau < 0. \tag{4.3}$$

Before moving into the consensus stability analysis, we need the following results.

Lemma 4.1.1 (**[26]**) *For a Laplacian matrix that satisfies Assumption 4.1.1, a similarity transformation T, with its first column $T_{(1)} = \mathbf{1}$ and the first row of T^{-1}, $T^{-1}_{(1)} = r^T$, can be constructed such that*

$$T^{-1}\mathcal{L}T = \overline{J}, \tag{4.4}$$

with $\overline{J}$ being a block diagonal matrix in the real Jordan form

$$\overline{J} = \begin{bmatrix} 0 & & & & & & \\ & \overline{J}_1 & & & & & \\ & & \ddots & & & & \\ & & & \overline{J}_p & & & \\ & & & & \overline{J}_{p+1} & & \\ & & & & & \ddots & \\ & & & & & & \overline{J}_q \end{bmatrix}, \tag{4.5}$$

where $\overline{J}_k \in \mathbb{R}^{n_k}$, $k = 1, 2, \ldots, p$, are the Jordan blocks for real eigenvalues $\lambda_k > 0$ with the multiplicity n_k in the form

$$\overline{J}_k = \begin{bmatrix} \lambda_k & 1 & & & \\ & \lambda_k & 1 & & \\ & & \ddots & \ddots & \\ & & & \lambda_k & 1 \\ & & & & \lambda_k \end{bmatrix},$$

and $\overline{J}_k \in \mathbb{R}^{2n_k}$, $k = p+1, p+2, \ldots, q$, are the Jordan blocks for conjugate eigenvalues $\alpha_k \pm j\beta_k$, $\alpha_k > 0$ and $\beta_k > 0$, with the multiplicity n_k in the form

$$\overline{J}_k = \begin{bmatrix} \nu(\alpha_k, \beta_k) & I_2 & & & \\ & \nu(\alpha_k, \beta_k) & I_2 & & \\ & & \ddots & \ddots & \\ & & & \nu(\alpha_k, \beta_k) & I_2 \\ & & & & \nu(\alpha_k, \beta_k) \end{bmatrix}$$

with I_2 being the identity matrix in $\mathbb{R}^{2\times 2}$ and

$$\nu(\alpha_k, \beta_k) = \begin{bmatrix} \alpha_i & \beta_i \\ -\beta_i & \alpha_i \end{bmatrix} \in \mathbb{R}^{2\times 2}.$$

4.2 H_∞ Consensus Control

For the multi-agent system (4.1), we have

$$x_i(t) = e^{Ah} x_i(t-h) + \int_{t-h}^{t} e^{A(t-\tau)} \left(Bu_i(\tau - h) + D\omega_i\right) \mathrm{d}\tau.$$

The control input takes the structure

$$u_i(t) = Ke^{Ah} \sum_{j=1}^{N} q_{ij} \left(x_i(t) - x_j(t)\right) = Ke^{Ah} \sum_{j=1}^{N} l_{ij} x_j(t), \tag{4.6}$$

where $K \in \mathbb{R}^{m\times n}$ is a constant control gain to be designed later. Under control algorithm (4.6), the multi-agent systems (4.1) can be written as

$$\begin{aligned}\dot{x}_i(t) &= Ax_i(t) + BK\sum_{j=1}^{N} l_{ij}x_j + D\omega_i \\ &\quad - BK\sum_{j=1}^{N} l_{ij}\int_{t-h}^{t} e^{A(t-\tau)}\left(Bu_j(\tau-h) + D\omega_j\right)\mathrm{d}\tau.\end{aligned}$$

Let $x = \left[x_1^T, x_2^T, \cdots, x_N^T\right]^T$, $u = \left[u_1^T, u_2^T, \cdots, u_N^T\right]^T$. The closed-loop system is then described by

$$\dot{x}(t) = \left(I_N \otimes A + \mathcal{L}\otimes BK\right)x(t) + \left(\mathcal{L}\otimes BK\right)(d_1 + d_2) + (I_N \otimes D)\omega, \quad (4.7)$$

where

$$\begin{aligned}d_1 &= -\int_{t-h}^{t}\left(I_N \otimes e^{A(t-\tau)}B\right)u(\tau-h)\mathrm{d}\tau, \\ d_2 &= -\int_{t-h}^{t}\left(I_N \otimes e^{A(t-\tau)}D\right)\omega(t)\mathrm{d}\tau.\end{aligned}$$

From the state transformation (4.2), we have

$$\xi(t) = x(t) - \left(\left(\mathbf{1}r^T\right)\otimes I_n\right)x(t) = \left(U\otimes I_n\right)x(t), \quad (4.8)$$

where $\xi = \left[\xi_1^T, \xi_2^T, \cdots, \xi_N^T\right]^T$, $U = I_N - \mathbf{1}r^T$. Since $r^T\mathbf{1} = 1$, it can be shown that $U\mathbf{1} = 0$. Therefore the consensus of system (4.1) is achieved when $\lim_{t\to\infty}\xi(t) = 0$, as $\xi = 0$ implies that $x_1 = x_2 = \cdots = x_N$, due to the fact that the null space of U is span$\{\mathbf{1}\}$. The dynamics of ξ can then be derived as

$$\dot{\xi} = \left(I_N \otimes A + \mathcal{L}\otimes BK\right)\xi + \left(\mathcal{L}\otimes BK\right)(d_1 + d_2) + \left(U\otimes D\right)\omega, \quad (4.9)$$

where we have used $r^T\mathcal{L} = 0$.

To explore the structure of $\mathcal{L}$, we propose another state transformation

$$\eta = \left(T^{-1}\otimes I_n\right)\xi, \quad (4.10)$$

with $\eta = \left[\eta_1^T, \eta_2^T, \cdots, \eta_N^T\right]^T$. Then we have

$$\dot{\eta} = \left(I_N \otimes A + \overline{J}\otimes BK\right)\eta + \Delta_1(x) + \Delta_2(t) + \Omega(t), \quad (4.11)$$

where

$$\begin{aligned}\Delta_1 &= \left(T^{-1}\mathcal{L}\otimes BK\right)d_1, \\ \Delta_2 &= \left(T^{-1}\mathcal{L}\otimes BK\right)d_2, \\ \Omega &= \left(T^{-1}U\otimes D\right)\omega(t).\end{aligned}$$

From state transformations (4.8) and (4.10), we have

$$\eta_1 = \left(r^T \otimes I_n\right)\xi = \left((r^T U) \otimes I_n\right) x \equiv 0.$$

With the control law shown in (4.6), the control gain matrix K is chosen as

$$K = -B^T P, \tag{4.12}$$

where P is a positive definite matrix to be designed. In the remainder of the chapter, Lyapunov-function-based analysis will be carried out to identify a condition for P to ensure that the consensus problem is solved by using the control algorithm (4.12) with control gain K in (4.6).

The consensus analysis will be carried out in terms of η. Let

$$V_i = \eta_i^T P \eta_i, \tag{4.13}$$

for $i = 2, 3, \cdots, N$. Then, let

$$V_0 = \sum_{i=2}^{N} V_i.$$

For the convenience of presentation, we recall from [26] the following results on V_0.

Lemma 4.2.1 *For multi-agent systems (4.1) with the transformed state η, $\dot{V}_0$ has following bounds specified in one of the following two cases:*
(1) If the eigenvalues of the Laplacian matrix $\mathcal{L}$ are distinct, $\dot{V}_0$ satisfies

$$\begin{aligned}\dot{V}_0 \le& \sum_{i=2}^{N} \eta_i^T \left(A^T P + PA - 2\alpha PBB^T P + (\kappa_1 + \kappa_2)PP\right)\eta_i \\ &+ \frac{1}{\kappa_1}\|\Delta_1\|^2 + \frac{1}{\kappa_2}\|\Delta_2\|^2 + 2\eta^T (I_N \otimes P)\Omega,\end{aligned} \tag{4.14}$$

where κ_1 and κ_2 are any positive real numbers, and

$$\alpha = \min\{\lambda_1, \lambda_2, \ldots, \lambda_p, \alpha_{p+1}, \alpha_{p+2}, \ldots, \alpha_q\}.$$

(2) If the Laplacian matrix $\mathcal{L}$ has multiple eigenvalues, $\dot{V}_0$ satisfies

$$\begin{aligned}\dot{V}_0 \le& \sum_{i=2}^{N} \eta_i^T \left(A^T P + PA - 2(\alpha - 1)PBB^T P + (\kappa_1 + \kappa_2)PP\right)\eta_i \\ &+ \frac{1}{\kappa_1}\|\Delta_1\|^2 + \frac{1}{\kappa_2}\|\Delta_2\|^2 + 2\eta^T (I_N \otimes P)\Omega,\end{aligned} \tag{4.15}$$

where κ_1 and κ_2 are any positive real numbers.

The following lemmas give the bounds of $\|\Delta_1\|^2$ and $\|\Delta_2\|^2$.

Lemma 4.2.2 *For the term $\Delta_1(x)$ shown in the transformed system dynamics (4.11), a bound can be established as*

$$\|\Delta_1\|^2 \leq \rho_1 \int_{t-h}^{t} \eta^T(\tau-h)\,\eta(\tau-h)\,\mathrm{d}\tau, \tag{4.16}$$

where

$$\rho_1 = 4ha_1^4 e^{2\lambda h} \lambda_\sigma^2\left(T^{-1}\right) \|\mathcal{L}\|_F^2 \|Q\|_F^2 \|T\|_F^2,$$

with a_1 and λ being positive numbers such that

$$a_1^2 I \geq PBB^TBB^TP, \tag{4.17}$$

$$\lambda I > A^T + A, \tag{4.18}$$

and $\lambda_\sigma(\cdot)$ and $\|\cdot\|_F$ being the maximum singular value and the Frobenius norm of a matrix, respectively.

Proof 11 *By the definition of $\Delta_1(x)$ in (4.11), we have*

$$\|\Delta_1\| = \left\|\left(T^{-1}\otimes I_n\right)(\mathcal{L}\otimes BK)d_1\right\| \leq \lambda_\sigma\left(T^{-1}\right)\|\mu\|, \tag{4.19}$$

where $\mu = \left[\mu_1^T, \mu_2^T, \ldots, \mu_N^T\right]^T = (\mathcal{L}\otimes BK)d_1$. Then, from (4.7) and (4.12), we have

$$\mu_i = BB^TP\sum_{j=1}^{N} l_{ij}\int_{t-h}^{t} e^{A(t-\tau)}BB^TPe^{Ah}\sum_{k=1}^{N} q_{jk}\left(x_k(\tau-h)-x_j(\tau-h)\right)\mathrm{d}\tau. \tag{4.20}$$

From $\eta = \left(T^{-1}\otimes I_n\right)\xi$, we obtain $\xi = (T\otimes I_n)\,\eta$. Then, from the state transformation (4.8), we have

$$\begin{aligned} x_k(t) - x_j(t) &= \xi_k(t) - \xi_j(t) \\ &= \left((T_k - T_j)\otimes I_n\right)\eta(t) \\ &= \sum_{l=1}^{N}\left(T_{kl} - T_{jl}\right)\eta_l(t), \end{aligned} \tag{4.21}$$

where T_k denotes the kth row of T. Define

$$\sigma_l = BB^TP\int_{t-h}^{t} e^{A(t-\tau)}BB^TPe^{Ah}\eta_l(\tau-h)\mathrm{d}\tau. \tag{4.22}$$

Then, from (4.20) and (4.21), we can obtain that

$$\mu_i = \sum_{j=1}^{N} l_{ij}\sum_{k=1}^{N} q_{jk}\sum_{l=1}^{N}\left(T_{kl}-T_{jl}\right)\sigma_l.$$

For the notational convenience, let $\sigma = \left[\sigma_1^T, \sigma_2^T, \ldots, \sigma_N^T\right]^T$. *It then follows that*

$$\begin{aligned}
\|\mu_i\| &\leq \sum_{j=1}^{N} |l_{ij}| \sum_{k=1}^{N} |q_{jk}| \, \|T_k\| \, \|\sigma\| + \sum_{k=1}^{N}\sum_{j=1}^{N} |l_{ij}| \, |q_{jk}| \, \|T_j\| \, \|\sigma\| \\
&\leq \sum_{j=1}^{N} |l_{ij}| \, \|q_j\| \, \|T\|_F \, \|\sigma\| + \sum_{k=1}^{N}\sum_{j=1}^{N} |l_{ij}| \, \|q_k\| \, \|T\|_F \, \|\sigma\| \\
&\leq 2 \|l_i\| \, \|Q\|_F \, \|T\|_F \, \|\sigma\| ,
\end{aligned} \tag{4.23}$$

where l_i *denotes the ith row of* $\mathcal{L}$. *Therefore, we have*

$$\|\mu\|^2 = \sum_{i=1}^{N} \|\mu_i\|^2 \leq 4 \|\mathcal{L}\|_F^2 \, \|Q\|_F^2 \, \|T\|_F^2 \, \|\sigma\|^2 , \tag{4.24}$$

where we have used $\sum_{i=1}^{N} \|l_i\|^2 = (\|\mathcal{L}\|_F)^2$.

Next we need to deal with $\|\sigma\|^2$. *By Lemma 2.1.2, we have*

$$\begin{aligned}
\|\sigma_i\|^2 &\leq h \int_{t-h}^{t} \eta_i^T(\tau - h) e^{A^T h} PBB^T e^{A^T(t-\tau)} PBB^T \\
&\quad \times BB^T P e^{A(t-\tau)} BB^T P e^{Ah} \eta_i(\tau - h) \mathrm{d}\tau \\
&\leq h a_1^2 \int_{t-h}^{t} \eta_i^T(\tau - h) e^{A^T h} PBB^T e^{A^T(t-\tau)} \\
&\quad \times e^{A(t-\tau)} BB^T P e^{Ah} \eta_i(\tau - h) \mathrm{d}\tau,
\end{aligned}$$

where a_1 *is a positive real number such that*

$$a_1^2 I \geq PBB^T BB^T P.$$

In view of Lemma 2.1.1 with $P = I$, *provided that*

$$R = -A^T - A + \lambda I > 0,$$

we have

$$e^{A^T t} e^{At} < e^{\lambda t} I,$$

and

$$\begin{aligned}
\|\sigma_i\|^2 &\leq h a_1^2 \int_{t-h}^{t} e^{\lambda(t-\tau)} \eta_i^T(\tau - h) e^{A^T h} PBB^T BB^T P e^{Ah} \eta_i(\tau - h) \mathrm{d}\tau \\
&\leq h a_1^4 e^{2\lambda h} \int_{t-h}^{t} \eta_i^T(\tau - h) \eta_i(\tau - h) \mathrm{d}\tau.
\end{aligned}$$

Then, $\|\sigma\|^2$ *can be bounded as*

$$\|\sigma\|^2 = \sum_{i=1}^{N} \|\sigma_i\|^2 \leq h a_1^4 e^{2\lambda h} \int_{t-h}^{t} \eta^T(\tau - h) \eta(\tau - h) \mathrm{d}\tau. \tag{4.25}$$

Hence, together with (4.19), (4.24) and (4.25), we get

$$\|\Delta_1\|^2 \leq \rho_1 \int_{t-h}^{t} \eta^T(\tau-h)\,\eta(\tau-h)\,\mathrm{d}\tau.$$

This completes the proof.

Lemma 4.2.3 *For the term $\Delta_2(t)$ in the transformed system dynamics (4.11), a bound can be established as*

$$\|\Delta_2\|^2 \leq \rho_2 \int_{t-h}^{t} \omega^T(\tau)\omega(\tau)\mathrm{d}\tau, \tag{4.26}$$

where

$$\rho_2 = ha_1^2 a_2 e^{\lambda h} \lambda_\sigma^2 \left(T^{-1}\right) \|\mathcal{L}\|_F^2,$$

with a_1 and λ being as defined in (4.17) and (4.18), and a_2 is a positive real number such that

$$a_2 I \geq D^T D. \tag{4.27}$$

Proof 12 *In a way similar to Lemma 3, we have*

$$\|\Delta_2\| = \left\|\left(T^{-1}\otimes I_n\right)\left(\mathcal{L}\otimes BK\right)d_2\right\| \leq \lambda_\sigma\left(T^{-1}\right)\|z\|, \tag{4.28}$$

where $z = (\mathcal{L}\otimes BK)d_2$. Let $z = \left[z_1^T, z_2^T, \ldots, z_N^T\right]^T$. Then from (4.7) and (4.12), we have

$$z_i = \sum_{j=1}^{N} l_{ij} BB^T P \int_{t-h}^{t} e^{A(t-\tau)} D\omega_j \mathrm{d}\tau.$$

It follows that

$$\|z_i\|^2 = \sum_{j=1}^{N} l_{ij}^2 \int_{t-h}^{t} \omega_j^T D^T e^{A^T(t-\tau)}\mathrm{d}\tau PBB^TBB^TP \int_{t-h}^{t} e^{A(t-\tau)} D\omega_j \mathrm{d}\tau.$$

With Lemma 2.1.2 and the condition (4.17), we have

$$\|z_i\|^2 \leq ha_1^2 \sum_{j=1}^{N} l_{ij}^2 \int_{t-h}^{t} \omega_j^T D^T e^{A^T(t-\tau)} e^{A(t-\tau)} D\omega_j \mathrm{d}\tau.$$

In view of Lemma 2.1.1, with the conditions (4.18) and (4.27), we have

$$\|z_i\|^2 \leq ha_1^2 a_2 e^{\lambda h} \int_{t-h}^{t} \sum_{j=1}^{N} l_{ij}^2 \|\omega_j\|^2 \mathrm{d}\tau.$$

Consequently,

$$\begin{aligned}\|z\|^2 &\le ha_1^2a_2e^{\lambda h}\int_{t-h}^{t}\sum_{i=1}^{N}\sum_{j=1}^{N} l_{ij}^2\,\|\omega_j\|^2\,\mathrm{d}\tau \\ &\le ha_1^2a_2e^{\lambda h}\|L\|_F^2\int_{t-h}^{t}\omega^T(\tau)\omega(\tau)\mathrm{d}\tau. \end{aligned}\tag{4.29}$$

Putting (4.28) and (4.29) together, we have

$$\|\Delta_2\|^2 \le \rho_2\int_{t-h}^{t}\omega^T(\tau)\omega(\tau)\mathrm{d}\tau.$$

This completes the proof.

For the first integral term shown in (4.16), we consider the following Krasovskii functional

$$W_1 = e^h\int_{t-h}^{t}\eta^T(\tau)\eta(\tau)\mathrm{d}\tau + e^h\int_{t-h}^{t}e^{\tau-t}\eta^T(\tau-h)\eta(\tau-h)\mathrm{d}\tau.$$

A direct evaluation gives that

$$\begin{aligned}\dot{W}_1 =& -e^h\int_{t-h}^{t}e^{\tau-t}\eta^T(\tau-h)\eta(\tau-h)\mathrm{d}\tau - \eta^T(t-2h)\eta(t-2h) + e^h\eta^T(t)\eta(t) \\ \le& -\int_{t-h}^{t}\eta^T(\tau-h)\eta(\tau-h)\mathrm{d}\tau + e^h\eta(t)^T\eta(t). \end{aligned}\tag{4.30}$$

For the second integral term shown in (4.26), we consider the following Krasovskii functional

$$W_2 = e^h\int_{t-h}^{t}e^{\tau-t}\omega^T(\tau)\omega(\tau)\mathrm{d}\tau.$$

A direct evaluation gives that

$$\begin{aligned}\dot{W}_2 =& -e^h\int_{t-h}^{t}e^{\tau-t}\omega^T(\tau)\omega(\tau)\mathrm{d}\tau + e^h\omega^T(\tau)\omega(\tau) - \omega^T(t-h)\omega(t-h) \\ \le& -\int_{t-h}^{t}\omega^T(\tau)\omega(\tau)\mathrm{d}\tau + e^h\omega^T(\tau)\omega(\tau). \end{aligned}\tag{4.31}$$

Let

$$V = V_0 + \frac{\rho_1}{\kappa_1}W_1 + \frac{\rho_2}{\kappa_2}W_2. \tag{4.32}$$

From (4.14), (4.15), (4.30) and (4.31), we obtain that

$$\dot{V} \le \eta^T(t)\left[I_N\otimes\left(H + \frac{\rho_1}{\kappa_1}e^hI_n\right)\right]\eta(t) + \frac{\rho_2}{\kappa_2}e^h\omega^T(t)\omega(t) + 2\eta^T(I_N\otimes P)\Omega, \tag{4.33}$$

where

$$H := A^T P + PA - 2\alpha PBB^T P + (\kappa_1 + \kappa_2)PP,$$

for Case (1), and

$$H := A^T P + PA - 2(\alpha - 1)PBB^T P + (\kappa_1 + \kappa_2)PP,$$

for Case (2).

The above expressions can be used for the H_∞ consensus analysis. The following theorem summarizes the results.

Theorem 8 *For an input-delayed multi-agent system (4.1) with the associated Laplacian matrix that satisfies Assumption 4.1.1, the H_∞ consensus control problem can be solved by the control algorithm (4.6) with the control gain $K = -B^T P$ specified in one of the following two cases:*

(1) If the eigenvalues of the Laplacian matrix $\mathcal{L}$ are distinct, the consensus is achieved if the following conditions are satisfied for $W = P^{-1}$, $a_1 > 0$ and $\lambda \geq 0$,

$$(A - \frac{1}{2}\lambda I_n)^T + (A - \frac{1}{2}\lambda I_n) < 0, \tag{4.34}$$

$$a_1 W \geq BB^T, \tag{4.35}$$

$$\begin{bmatrix} \Gamma_1 & W & D \\ W & -\left(\frac{\rho_1 e^h}{\kappa_1} + a_3\right)^{-1} I_n & 0 \\ D^T & 0 & -\left(\gamma^2 - \frac{\rho_2}{\kappa_2} e^h\right) a_4^{-1} \end{bmatrix} < 0, \tag{4.36}$$

where $\Gamma_1 = WA^T + AW - 2\alpha BB^T + (\kappa_1 + \kappa_2) I_n$, $\kappa_1 > 0$, $\kappa_2 > \frac{\rho_2}{\gamma^2} e^h$, $a_3 \geq \lambda_{\max}(T^T T \otimes C^T C)$ and $a_4 \geq \lambda_{\max}(T^{-1} U U^T (T^{-1})^T)$.

(2) If the Laplacian matrix $\mathcal{L}$ has multiple eigenvalues, the consensus is achieved if the conditions (4.34), (4.35) and the following condition are satisfied for $W = P^{-1}$, $a_1 > 0$, $\lambda \geq 0$,

$$\begin{bmatrix} \Gamma_2 & W & D \\ W & -\left(\frac{\rho_1 e^h}{\kappa_1} + a_3\right)^{-1} I_n & 0 \\ D^T & 0 & -\left(\gamma^2 - \frac{\rho_2}{\kappa_2} e^h\right) a_4^{-1} \end{bmatrix} < 0, \tag{4.37}$$

where $\Gamma_2 = WA^T + AW - 2(\alpha - 1)BB^T + (\kappa_1 + \kappa_2) I_n$.

Proof 13 *From (4.10), we obtain that*

$$e(t) = (I_N \otimes C)\xi = (I_N \otimes C)(T \otimes I_n)\eta(t).$$

It follows that

$$e^T(t)e(t) = \eta^T(t)(T^TT \otimes C^TC)\eta(t) \leq a_3\eta^T(t)\eta(t). \tag{4.38}$$

Under the zero-initial condition, $x(0) = 0$. *It is clear that* $V(0) = 0$. *Next, for any non-zero* ω, *we have*

$$\begin{aligned} J &= \int_0^\infty \left[e^T(t)e(t) - \gamma^2\omega^T(t)\omega(t) + \dot{V}\right] \mathrm{d}\tau - V(\infty) + V(0) \\ &\leq \int_0^\infty \eta^T \left[I_N \otimes \left(H + \frac{\rho_1}{\kappa_1}e^h I_n + a_3 I_n\right)\right] \eta \mathrm{d}\tau \\ &\quad + \int_0^\infty \left(\frac{\rho_2}{\kappa_2}e^h - \gamma^2\right) \omega^T(t)\omega(t) + 2\eta^T(I_N \otimes P)\Omega \mathrm{d}\tau \\ &= \int_0^\infty \begin{bmatrix} \eta \\ \omega \end{bmatrix}^T \Theta \begin{bmatrix} \eta \\ \omega \end{bmatrix} \mathrm{d}\tau, \end{aligned} \tag{4.39}$$

where

$$\Theta = \begin{bmatrix} I_N \otimes \left(H + \frac{\rho_1}{\kappa_1}e^h I_n + a_3 I_n\right) & T^{-1}U \otimes PD \\ \left(T^{-1}U\right)^T \otimes D^T P & \left(\frac{\rho_2}{\kappa_2}e^h - \gamma^2\right) I \end{bmatrix}.$$

Thus, $J < 0$ *if* $\Theta < 0$. *By Schur complement lemma, we know that* $\Theta < 0$ *if the following inequality hold*

$$H + \frac{\rho_1}{\kappa_1}e^h I_n + a_3 I_n + a_4\left(\gamma^2 - \frac{\rho_2}{\kappa_2}e^h\right)^{-1} PDD^TP < 0. \tag{4.40}$$

From $W = P^{-1}$, *and condition (4.33), it is obtained that (4.40) is equivalent to*

$$\begin{aligned} &WA^T + AW - 2\alpha BB^T + (\kappa_1 + \kappa_2)I_n \\ &\quad + \left(\frac{\rho_1}{\kappa_1}e^h + a_3\right) WW + a_4\left(\gamma^2 - \frac{\rho_2}{\kappa_2}e^h\right)^{-1} DD^T < 0, \end{aligned} \tag{4.41}$$

for Case (1), and

$$\begin{aligned} &WA^T + AW - 2(\alpha - 1)BB^T + (\kappa_1 + \kappa_2)I_n \\ &\quad + \left(\frac{\rho_1}{\kappa_1}e^h + a_3\right) WW + a_4\left(\gamma^2 - \frac{\rho_2}{\kappa_2}e^h\right)^{-1} DD^T < 0, \end{aligned} \tag{4.42}$$

for Case (2).

By Schur complement lemma, we know that the conditions (4.41) and (4.42) are equivalent to the conditions specified in (4.36) and (4.37). Considering conditions (4.34)–(4.37), we can obtain that $J < 0$. *Therefore, the* H_∞ *consensus problem is solved.*

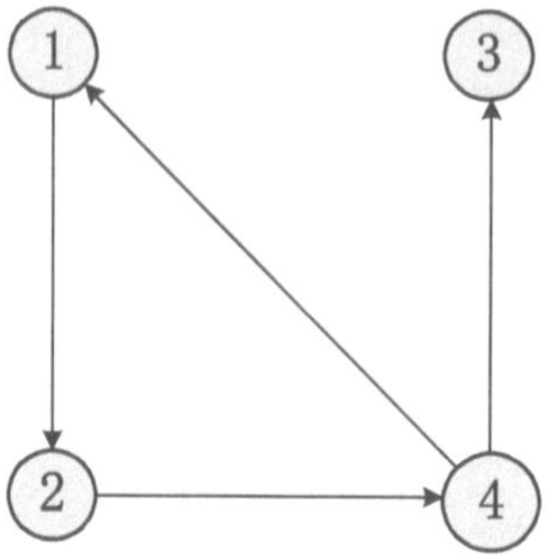

FIGURE 4.1: Communication topology.

The consensus analysis for multi-agent systems with directed graph can clearly be applied to the systems with undirected graphs. Indeed, it can be treated as a special situation of the Case 1. A corollary is given for this special case.

Corollary 4.2.1 *For an input-delayed multi-agent system (4.1) with undirected graph, the H_∞ consensus control problem can be solved by the control algorithm (4.6) with the control gain $K = -B^T P$ where P is a positive definite matrix satisfying conditions (4.34)–(4.36).*

4.3 A Numerical Example

In this section, a simulation study is carried out to demonstrate the effectiveness of the proposed control design. Consider a connection of four agents as shown in Figure 4.1. The dynamics of each agent is described by (4.1), with

$$A = \begin{bmatrix} 0 & -0.1 \\ 0.1 & 0.1 \end{bmatrix},\ B = \begin{bmatrix} 0.5 & 0 \\ 0 & 0.5 \end{bmatrix},\ C = \begin{bmatrix} 0.1 \\ 0 \end{bmatrix}^T,\ D = \begin{bmatrix} 0.1 & 0 \\ 0 & 0.1 \end{bmatrix}.$$

Note that A has two positive eigenvalues. The external disturbances $\omega = [2w, w, -2w, 1.5w]^T$, where $w(t)$ is a ten-period square wave starting at $t = 0$ with the width 5 and height 1. The input delay of the system is 0.1 s. The Laplacian matrix is given by

$$\mathcal{L} = \begin{bmatrix} 1 & 0 & 0 & -1 \\ -1 & 1 & 0 & 0 \\ 0 & 0 & 1 & -1 \\ 0 & -1 & 0 & 1 \end{bmatrix}.$$

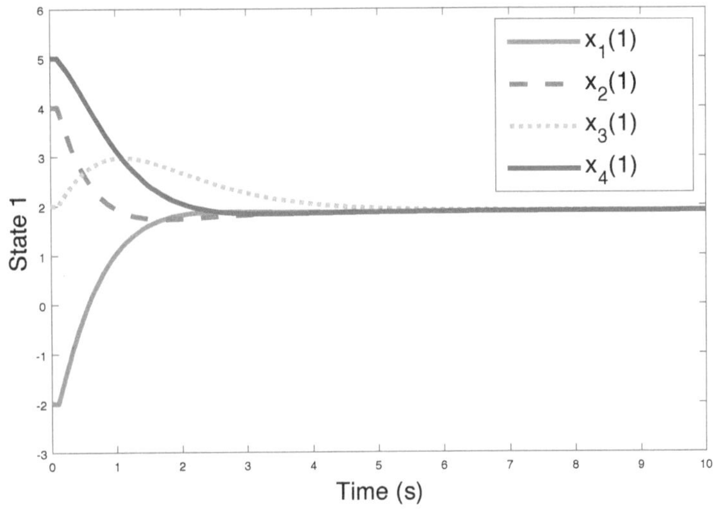

FIGURE 4.2: The state 1 of agents with $h = 0.1$.

The eigenvalues of $\mathcal{L}$ are $\{\ 0,\ 1,\ 3/2 \pm j\sqrt{3}/2\ \}$. Therefore, Assumption 4.1.1 is satisfied. We obtain that $\alpha = 1$ and $r^T = \left[\ \frac{1}{3},\ \frac{1}{3},\ 0,\ \frac{1}{3}\ \right]$. In this case, we choose the H_∞ performance index $\gamma = 1$, and the initial states of agents are chosen randomly within $[-5, 5]$, $u(\theta) = [0,\ 0,\ 0,\ 0]^T$, for $\theta \in [-h, 0]$. With the values of $\lambda = 0.2, a_1 = 0.3$, and $\kappa_1 = \kappa_2 = 0.1$, a feasible solution of the feedback gain K is found to be

$$K = \begin{bmatrix} -1.5349 & -0.1504 \\ -0.1504 & -1.5541 \end{bmatrix}.$$

Figures 4.2 and 4.3 show the simulation results for the state of each agent under the case $\omega = 0$. Clearly the conditions specified in Theorem 8 are sufficient for the control gain to achieve consensus control. Figure 4.4 shows the trajectories of the performance variables, $e_i(t), i = 1, \cdots, 4$, under the zero-initial condition. In addition, with the same control gain, the consensus control is still achieved for the multi-agent system with a much larger delay $h = 0.5$, as shown in Figures 4.5 and 4.6, which implies the conditions could be conservative in the control gain design for a given input delay.

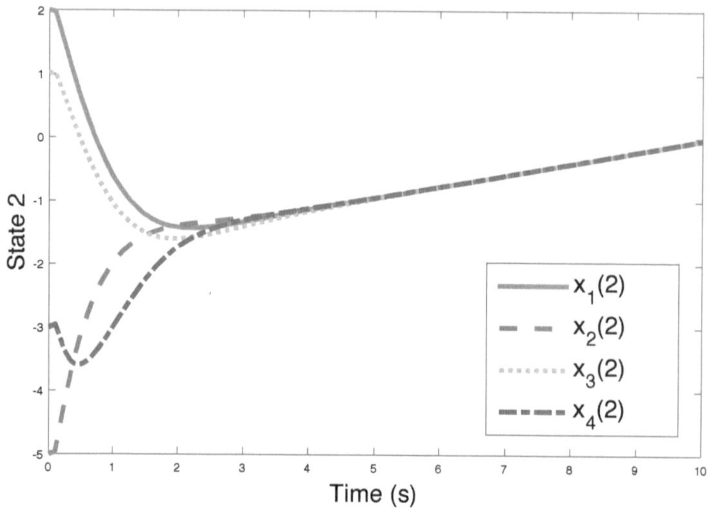

FIGURE 4.3: The state 2 of agents with $h = 0.1$.

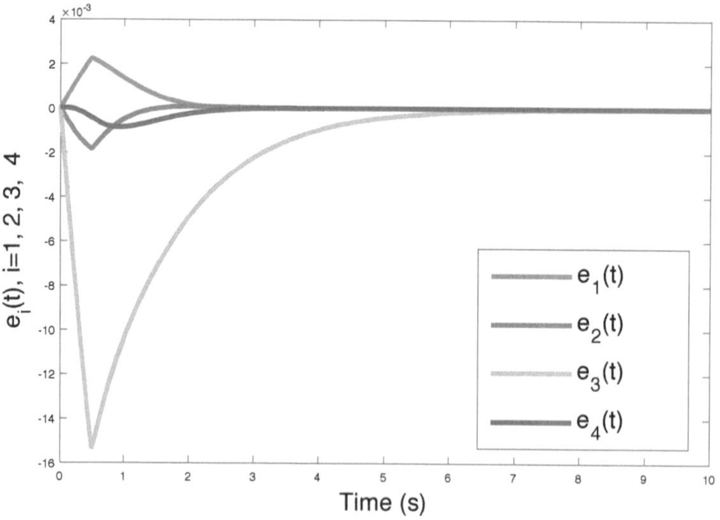

FIGURE 4.4: The trajectories of performance variables.

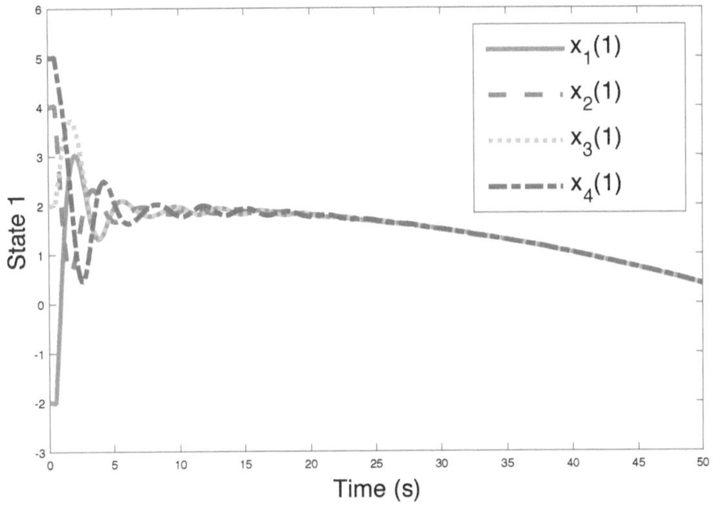

FIGURE 4.5: The state 1 of agents with $h = 0.5$.

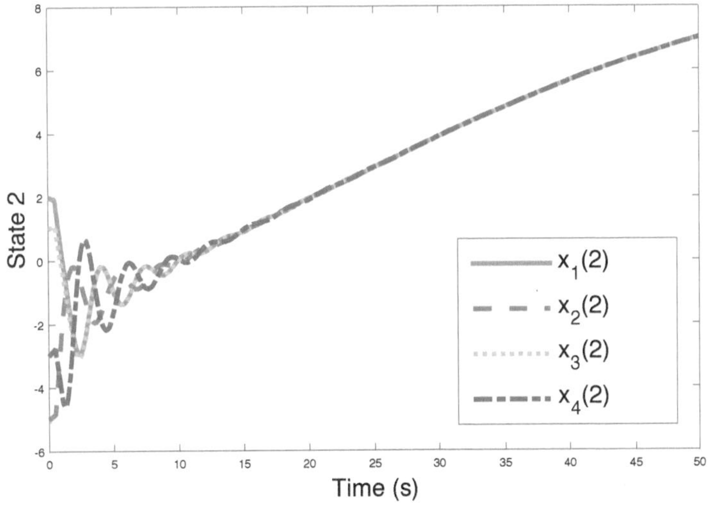

FIGURE 4.6: The state 2 of agents with $h = 0.5$.

4.4 Conclusions

In this chapter, we have addressed the H_∞ consensus problem for linear multi-agent systems with input delay and general directed graph. This input delay may represent some delays in the network communication. The truncated prediction feedback method is employed to deal with the input delay, and the integral terms that remain in the transformed systems are carefully analyzed by using Krasovskii functionals. By using the real Jordan form of the Laplacian matrix, sufficient conditions for the H_∞ consensus are identified through Lyapunov analysis. The conditions can be solved by employing LMIs with a set of iterative parameters. This chapter only requires the connection graph to have a spanning tree. Future work will focus on H_∞ consensus protocol design for multi-agent systems with time-varying input delay and Lipschitz nonlinearities.

4.5 Notes

This chapter extends an idea in [26, 30] and develops consensus protocols for general linear multi-agent systems with input delay and directed topology. The main materials of this chapter are based on [155]. It is worth mentioning that leaderless topology is considered in this chapter. The results in this chapter can be extended to leader-follower cases by following the procedures shown in [163, 164].

Chapter 5

Consensus Control of Nonlinear Multi-Agent Systems with Input Delay

This chapter deals with the consensus control problem for Lipschitz nonlinear multi-agent systems with input delay. First, predictor feedback approach is applied to compensate the input delay. Then, a prediction of the agent state over the delay period is approximated by the zero input solution of the agent dynamics. The structure of a linear state feedback control algorithm is assumed for each agent based on such approximated state prediction. By transforming the Laplacian matrix into the real Jordan form, sufficient conditions are established under which the multi-agent systems under the proposed control algorithms achieve global consensus. The feedback gain is then designed by solving these conditions with an iterative LMI procedure. Simulation studies are given to validate the proposed control design.

5.1 Problem Formulation

In this chapter, we consider control design for a group of N agents, each represented by a nonlinear subsystem that is subject to input delay and Lipschitz nonlinearity,

$$\dot{x}_i(t) = Ax_i(t) + Bu_i(t-h) + \phi(x_i), \tag{5.1}$$

where for agent i, $i = 1, 2, \ldots, N$, $x_i \in \mathbb{R}^n$ is the state vector, $u_i \in \mathbb{R}^m$ is the control input vector, $A \in \mathbb{R}^{n\times n}$ and $B \in \mathbb{R}^{n\times m}$ are constant matrices with (A, B) being controllable, $h > 0$ is input delay, and the initial conditions $x_i(\theta)$, $\theta \in [-h, 0]$, are given and bounded, and $\phi : \mathbb{R}^n \to \mathbb{R}^n$, $\phi(0) = 0$, is a Lipschitz nonlinear function with a Lipschitz constant γ, i.e., for any two constant vectors $a, b \in \mathbb{R}^n$,

$$\|\phi(a) - \phi(b)\| \leq \gamma\|a - b\|.$$

The connections among the agents are specified by a directed graph satisfied with the following assumption.

Assumption 5.1.1 *The eigenvalue of the Laplacian matrix at 0 is a single eigenvalue.*

The consensus control problem considered in this chapter is to design a control strategy, using the relative state information, to ensure that all agents asymptotically converge to an identical trajectory.

Before moving into the main result, we present a few preliminary results which are useful for the consensus analysis later in the chapter. We first present an overview of the Artstein-Kwon-Pearson reduction method. Consider an input-delayed system

$$\dot{x}(t) = Ax(t) + Bu(t-h) + \phi(x), \tag{5.2}$$

with $\phi : \mathbb{R}^n \to \mathbb{R}^n$, $\phi(0) = 0$, being a Lipschitz nonlinear function, and the initial conditions $x(\theta)$, $\theta \in [-h, 0]$, being bounded. Let

$$z(t) = x(t) + \int_t^{t+h} \mathrm{e}^{A(t-\tau)} Bu(\tau - h) \mathrm{d}\tau, \tag{5.3}$$

Differentiating $z(t)$ against time yields

$$\begin{aligned} \dot{z}(t) &= Ax(t) + \phi(x) + \mathrm{e}^{-Ah} Bu(t) + A \int_t^{t+h} \mathrm{e}^{A(t-\tau)} Bu(\tau - h) \mathrm{d}\tau \\ &= Az(t) + Du(t) + \phi(x), \end{aligned} \tag{5.4}$$

where $D = \mathrm{e}^{-Ah}B$. When $\phi(x)$ is absent, system (5.4) is delay free.

Remark 5.1.1 *It is straightforward to verify that the controllability of (A, B) and $(A, \mathrm{e}^{-Ah}B)$ are equivalent. We need additional conditions in the sequel due to the nonlinearity and the consensus requirement.*

We consider a controller

$$u(t) = Kz(t). \tag{5.5}$$

From (5.3) and (5.5), we have

$$\|x(t)\| \leq \|z(t)\| + h \left(\max_{-h \leq \theta \leq 0} \|\mathrm{e}^{A\theta}\| \right) \|B\| \|K\| \|z_t(\theta)\|,$$

where $z_t(\theta) := z(t+\theta)$, $-h \leq \theta \leq 0$. Thus, $x(t) \to 0$ as $z(t) \to 0$. In other words, if the controller (5.5) stabilizes the transformed system (5.4), then the original system (5.2) is also stable with the same controller [66].

Remark 5.1.2 *With any given bounded initial condition $u(\theta)$, $\theta \in [-h, 0]$, a stable feedback controller* (5.5) *implies that $u(t)$ in* (5.3) *is locally integrable, which allows for the model reduction as* (5.4).

5.2 Predictor-Based Consensus

For the multi-agent system (5.1), we use (5.3) to transform the agent dynamics to

$$\dot{z}_i(t) = Az_i + Du_i(t) + \phi(x_i), \tag{5.6}$$

where $D = \mathrm{e}^{-Ah}B$.

We propose a control design using the relative state information. The control input takes the structure,

$$u_i = -K\sum_{j=1}^{N} l_{ij}z_j, \tag{5.7}$$

where $K \in \mathbb{R}^{m\times n}$ is a constant control gain matrix to be designed later.

The closed-loop system is then described by

$$\dot{z} = (I_N \otimes A - L \otimes DK)z + \Phi(x), \tag{5.8}$$

where

$$z = \begin{bmatrix} z_1 \\ z_2 \\ \vdots \\ z_N \end{bmatrix}, \quad \Phi(x) = \begin{bmatrix} \phi(x_1) \\ \phi(x_2) \\ \vdots \\ \phi(x_N) \end{bmatrix}.$$

For a Laplacian matrix that satisfies Assumption 5.1.1, there exists a similarity transformation T, with its first column being $T_1 = \mathbf{1}$, such that

$$T^{-1}\mathcal{L}T = J. \tag{5.9}$$

where J has the same form with that in Lemma 3.1.1. Define $r^T \in \mathbb{R}^{1\times N}$ as the left eigenvector of $\mathcal{L}$ corresponding to the eigenvalue at 0, that is, $r^T\mathcal{L} = 0$. Furthermore, let r be scaled such that $r^T\mathbf{1} = 1$ and let the first row of T^{-1} be $(T^{-1})_1 = r^T$.

Based on the vector r, we introduce a state transformation

$$\xi_i = z_i - \sum_{j=1}^{N} r_j z_j, \tag{5.10}$$

for $i = 1, 2\ldots, N$. Let

$$\xi = [\xi_1^T, \xi_2^T, \cdots, \xi_N^T]^T.$$

We have

$$\xi = z - ((\mathbf{1}r^T) \otimes I_n)z = (M \otimes I_n)z,$$

where $M = I_N - \mathbf{1}r^T$. Since $r^T\mathbf{1} = 1$, it can be shown that $M\mathbf{1} = 0$. Therefore the consensus of system (5.8) is achieved when $\lim_{t\to\infty} \xi(t) = 0$, as $\xi = 0$ implies $z_1 = z_2 = \cdots = z_N$, due to the fact that the null space of M is span($\mathbf{1}$). The dynamics of ξ can then be obtained as

$$\begin{aligned}\dot{\xi} &= (I_N \otimes A - \mathcal{L} \otimes DK)z - \mathbf{1}r^T \otimes I_N[I_N \otimes A - \mathcal{L} \otimes DK]z + (M \otimes I_n)\Phi(x)\\ &= (I_N \otimes A - \mathcal{L} \otimes DK)\xi + (M \otimes I_n)\Phi(x). \qquad (5.11)\end{aligned}$$

To explore the structure of $\mathcal{L}$, let us introduce another state transformation

$$\eta = (T^{-1} \otimes I_n)\xi. \qquad (5.12)$$

Then we have

$$\dot{\eta} = (I_N \otimes A - J \otimes DK)\eta + \Psi(x), \qquad (5.13)$$

where $\Psi(x) = (T^{-1}M \otimes I_n)\Phi(x)$, and

$$\eta = \begin{bmatrix} \eta_1 \\ \eta_2 \\ \vdots \\ \eta_N \end{bmatrix}, \Psi(x) = \begin{bmatrix} \psi_1(x) \\ \psi_2(x) \\ \vdots \\ \psi_N(x) \end{bmatrix},$$

with $\eta_i \in \mathbb{R}^n$ and $\psi_i : \mathbb{R}^{n\times N} \to \mathbb{R}^n$ for $i = 1, 2, \ldots, N$. Then from (5.10) and (5.12), we have:

$$\eta_1 = \left(r^T \otimes I_n\right)\xi = \left((r^T M) \otimes I_n\right) z \equiv 0.$$

The nonlinear term $\Psi(x)$ in the transformed system dynamic model (5.13) is expressed as a function of the state x. For the stability analysis, first we need to establish a bound of this nonlinear function in terms of the transformed state η. The following lemma gives a bound of $\Psi(x)$.

Lemma 5.2.1 *For the nonlinear term $\Psi(x)$ in the transformed system dynamics (5.13), a bound can be established in terms of the state η as*

$$\|\Psi\|^2 \le \gamma_0^2(\|\eta\|^2 + 4\lambda_\sigma^2(Q)\,\|\delta\|^2), \qquad (5.14)$$

with

$$\gamma_0 = 2\sqrt{2N}\gamma\,\|r\|\,\|T\|_F\,\lambda_\sigma(T^{-1}), \qquad (5.15)$$

$$\delta = -\int_t^{t+h} e^{A(t-\tau)}BK\eta(\tau - h)\mathrm{d}\tau, \qquad (5.16)$$

where $\lambda_\sigma(\cdot)$ and $\|\cdot\|_F$ denote the maximum singular value and Frobenius norm of a matrix, respectively.

Proof 14 *Based on the state transformations (5.10) and (5.12), we have*

$$\Psi(x) = \left(T^{-1} \otimes I_n\right)(M \otimes I_n)\,\Phi(x) = \left(T^{-1} \otimes I_n\right)\mu,$$

where $\mu = (M \otimes I_n)\Phi(x)$*. Then, we have*

$$\|\Psi(x)\| \leq \lambda_\sigma(T^{-1})\,\|\mu\|\,, \tag{5.17}$$

where $\mu = [\mu_1, \mu_2, \ldots, \mu_N]^T$.

Recalling that $M = I_N - \mathbf{1}r^T$*, we have*

$$\mu_i = \phi(x_i) - \sum_{k=1}^{N} r_k \phi(x_k) = \sum_{k=1}^{N} r_k\left(\phi(x_i) - \phi(x_k)\right).$$

It then follows that

$$\|\mu_i\| \leq \gamma \sum_{k=1}^{N} |r_k|\,\|x_i - x_k\| \tag{5.18}$$

From the state transformation (5.3), we have

$$x_i - x_k = (z_i - \sigma_i) - (z_k - \sigma_k) = (z_i - z_k) - (\sigma_i - \sigma_k)$$

where

$$\sigma_i = \int_t^{t+h} \mathrm{e}^{A(t-\tau)} B u_i(\tau - h)\mathrm{d}\tau.$$

Then, we have

$$\|\mu_i\| \leq \gamma \sum_{k=1}^{N} |r_k|\left(\|z_i - z_k\| + \|\sigma_i - \sigma_k\|\right). \tag{5.19}$$

From $\eta = (T^{-1} \otimes I_n)\xi$*, we obtain* $\xi = (T \otimes I_n)\eta$*, and from the state transformations (5.10), we have*

$$z_i - z_k = \xi_i - \xi_k = \left((t_i - t_k) \otimes I_n\right)\eta = \sum_{j=1}^{N} (t_{ij} - t_{kj})\eta_j,$$

where t_k *denotes the* k*th row of* T*. Then, we obtain*

$$\|z_i - z_k\| \leq \left(\|t_i\| + \|t_k\|\right)\|\eta\|\,. \tag{5.20}$$

We next deal with the derived terms σ_i *and* σ_k*. We have*

$$\begin{aligned}\sum_{k=1}^{N} |r_k|\,\|\sigma_i - \sigma_k\| &\leq \sum_{k=1}^{N} |r_k|\,\|\sigma_i\| + \sum_{k=1}^{N} |r_k|\,\|\sigma_k\| \\ &\leq \|r\|\sqrt{N}\,\|\sigma_i\| + \|r\|\,\|\sigma\|\,,\end{aligned} \tag{5.21}$$

where $\sigma = [\sigma_1^T, \sigma_2^T, \cdots, \sigma_N^T]^T$, *and we used the inequality*

$$\sum_{i=1}^{N} |a_i| \leq \sqrt{N} \|a\| .$$

Then, from (5.19), (5.20) and (5.21), we can obtain that

$$\begin{aligned}
\|\mu_i\| &\leq \gamma \sum_{k=1}^{N} |r_k| \left(\|t_i\| + \|t_k\|\right) \|\eta\| + \gamma\sqrt{N} \|r\| \|\sigma_i\| + \gamma \|r\| \|\sigma\| \\
&\leq \gamma(\|r\| \sqrt{N} \|t_i\| + \|r\| \|T\|_F) \|\eta\| + \gamma\sqrt{N} \|r\| \|\sigma_i\| + \gamma \|r\| \|\sigma\| \\
&= \gamma \|r\| [(\sqrt{N} \|t_i\| + \|T\|_F) \|\eta\| + \sqrt{N} \|\sigma_i\| + \|\sigma\|]. \qquad (5.22)
\end{aligned}$$

It then follows that

$$\begin{aligned}
\|\mu\|^2 &= \sum_{i=1}^{N} (\|\mu_i\|)^2 \leq (\sum_{i=1}^{N} \|\mu_i\|)^2 \\
&\leq 4\gamma^2 \|r\|^2 \sum_{i=1}^{N} (N \|t_i\|^2 + \|T\|_F^2) \|\eta\|^2 + 4\gamma^2 \|r\|^2 \sum_{i=1}^{N} (N \|\sigma_i\|^2 + \|\sigma\|^2) \\
&= 8\gamma^2 \|r\|^2 N \left[\|T\|_F^2 \|\eta\|^2 + \|\sigma\|^2\right], \qquad (5.23)
\end{aligned}$$

where we have used

$$\sum_{k=1}^{N} \|t_k\|^2 = \|T\|_F^2 ,$$

and the inequality

$$(a + b + c + d)^2 \leq 4(a^2 + b^2 + c^2 + d^2).$$

Next we need to deal with $\|\sigma\|^2$. *From (5.7), we can get*

$$\sigma_i = \int_t^{t+h} \mathrm{e}^{A(t-\tau)} B u_i(\tau - h) \mathrm{d}\tau = -\int_t^{t+h} \mathrm{e}^{A(t-\tau)} BK \sum_{j=1}^{N} l_{ij} z_j(\tau - h) \mathrm{d}\tau.$$

From the relationship between Q *and* L, *we have*

$$\begin{aligned}
\sum_{j=1}^{N} l_{ij} z_j &= \sum_{j=1}^{N} q_{ij} (z_i - z_j) \\
&= \sum_{j=1}^{N} q_{ij} \left((t_i - t_j) \otimes I_n\right) \eta \\
&= \sum_{j=1}^{N} q_{ij} \sum_{l=1}^{N} (t_{il} - t_{jl}) \eta_l. \qquad (5.24)
\end{aligned}$$

Here we define δ_l

$$\delta_l = -\int_t^{t+h} \mathrm{e}^{A(t-\tau)} BK\eta_l(\tau - h)\mathrm{d}\tau. \tag{5.25}$$

Then we can obtain that

$$\sigma_i = \sum_{j=1}^{N} q_{ij} \sum_{l=1}^{N} (t_{il} - t_{jl})\delta_l.$$

It then follows that

$$\|\sigma_i\| \le \sum_{j=1}^{N} q_{ij} \left(\|t_i\| + \|t_j\|\right) \|\delta\| . \tag{5.26}$$

where $\delta = [\delta_1^T, \delta_2^T, \cdots, \delta_N^T]^T$. *With (5.26), the sum of the* $\|\sigma_i\|$ *can be obtained*

$$\begin{aligned}
\sum_{i=1}^{N} \|\sigma_i\| &\le \|\delta\| \sum_{i=1}^{N}\sum_{j=1}^{N} q_{ij}(\|t_i\| + \|t_j\|) \\
&= \|\delta\| \sum_{i=1}^{N}\sum_{j=1}^{N} q_{ij} \|t_i\| + \|\delta\| \sum_{i=1}^{N}\sum_{j=1}^{N} q_{ij} \|t_j\| \\
&\le \lambda_\sigma(Q) \|T\|_F \|\delta\| + \lambda_\sigma(Q^T) \|T\|_F \|\delta\| \\
&= 2\lambda_\sigma(Q) \|T\|_F \|\delta\| ,
\end{aligned} \tag{5.27}$$

with $\lambda_\sigma(Q) = \lambda_\sigma(Q^T)$. *Therefore we have*

$$\|\sigma\|^2 = \sum_{i=1}^{N} (\|\sigma_i\|)^2 \le \left(\sum_{i=1}^{N} \|\sigma_i\|\right)^2 \le 4\lambda_\sigma^2(Q) \|T\|_F^2 \|\delta\|^2 . \tag{5.28}$$

Hence, together with (5.23) and (5.28), we get

$$\|\mu\|^2 \le 8\gamma^2 \|r\|^2 N \|T\|_F^2 \left(\|\eta\|^2 + 4\lambda_\sigma^2(Q) \|\delta\|^2\right). \tag{5.29}$$

Finally, we obtain the bound for Ψ *as*

$$\|\Psi\|^2 \le \lambda_\sigma^2(T^{-1}) \|\mu\|^2 \le \gamma_0^2 \left(\|\eta\|^2 + 4\lambda_\sigma^2(Q) \|\delta\|^2\right), \tag{5.30}$$

with

$$\begin{aligned}
\gamma_0 &= 2\sqrt{2N}\gamma \|r\| \|T\|_F \lambda_\sigma(T^{-1}), \\
\delta &= -\int_t^{t+h} \mathrm{e}^{A(t-\tau)} BK\eta(\tau - h)\mathrm{d}\tau.
\end{aligned}$$

This completes the proof.

With the control law shown in (5.7), the control gain matrix K is chosen as

$$K = D^T P, \tag{5.31}$$

where P is a positive definite matrix. In the remaining part of the chapter, we will use Lyapunov-function-based analysis to identify a condition for P to ensure that consensus is achieved by using the control algorithm (5.7) with the control gain K in (5.31).

The stability analysis will be carried out in terms of η. As discussed earlier, the consensus control can be guaranteed by showing that η converges to zero, which is sufficed by showing that η_i converges to zero for $i = 2, 3, \ldots, N$, since we have shown that $\eta_1 = 0$.

From the structure of the Laplacian matrix shown in Lemma 3.1.1, we can see that

$$N_k = 1 + \sum_{j=1}^{k} n_j,$$

for $k = 1, 2, \ldots, q$. Note that $N_q = N$.

The agent state variables η_i from $i = 2$ to N_p are the state variables which are associated with the Jordan blocks of real eigenvalues, and η_i for $i = N_p+1$ to N are with Jordan blocks of complex eigenvalues.

For the state variables associated with the Jordan blocks J_k of real eigenvalues, i.e., for $k \leq p$, we have the dynamics given by

$$\dot{\eta}_i = (A - \lambda_k DD^T P)\eta_i - DD^T P\eta_{i+1} + \psi_i(x),$$

for $i = N_{k-1} + 1,\ N_{k-1} + 2, \cdots, N_k - 1$, and

$$\dot{\eta}_i = (A - \lambda_k DD^T P)\eta_i + \psi_i(x),$$

for $i = N_k$.

For the state variables associated with the Jordan blocks J_k, i.e., for $k > p$, corresponding to complex eigenvalues, we consider the dynamics of the state variables in pairs. For notational convenience, let

$$\begin{aligned} i_1(j) &= N_{k-1} + 2j - 1 \\ i_2(j) &= N_{k-1} + 2j \end{aligned}$$

for $j = 1, 2, \ldots, n_k/2$. The dynamics of η_{i_1} and η_{i_2} for $j = 1, 2, \ldots, n_k/2 - 1$ are expressed by

$$\begin{aligned} \dot{\eta}_{i_1} &= (A - \alpha_k DD^T P)\eta_{i_1} - \beta_k DD^T P\eta_{i_2} - DD^T P\eta_{i_1+2} + \psi_{i_1}, \\ \dot{\eta}_{i_2} &= (A - \alpha_k DD^T P)\eta_{i_2} + \beta_k DD^T P\eta_{i_1} - DD^T P\eta_{i_2+2} + \psi_{i_2}. \end{aligned}$$

For $j = n_k/2$, we have

$$\begin{aligned} \dot{\eta}_{i_1} &= (A - \alpha_k DD^T P)\eta_{i_1} - \beta_k DD^T P\eta_{i_2} + \psi_{i_1}, \\ \dot{\eta}_{i_2} &= (A - \alpha_k DD^T P)\eta_{i_2} + \beta_k DD^T P\eta_{i_1} + \psi_{i_2}. \end{aligned}$$

Let

$$V_i = \eta_i^T P \eta_i, \tag{5.32}$$

for $i = 2, 3 \ldots, N$. Let

$$V_0 = \sum_{i=2}^{N} \eta_i^T P \eta_i. \tag{5.33}$$

For the convenience of presentation, we borrow the following results for V_0 from [26].

Lemma 5.2.2 *For a network-connected dynamic system (5.1) with the transformed state η, $\dot{V}_0$ has following bounds specified in one of the following two cases:*
(1) If the eigenvalues of the Laplacian matrix $\mathcal{L}$ are distinct, i.e., $n_k = 1$ for $k = 1, 2, \ldots, q$, $\dot{V}_0$ satisfies

$$\dot{V}_0 \leq \sum_{i=2}^{N} \eta_i^T \left(A^T P + PA - 2\alpha PDD^T P + \kappa PP\right) \eta_i + \frac{1}{\kappa}\|\Psi\|^2, \tag{5.34}$$

with κ being any positive real number and

$$\alpha = \min\{\lambda_1, \lambda_2, \ldots, \lambda_p, \alpha_{p+1}, \alpha_{p+2}, \ldots, \alpha_q\}.$$

(2) If the Laplacian matrix $\mathcal{L}$ has multiple eigenvalues, i.e., $n_k > 1$ for any $k \in \{1, 2, \cdots, q\}$, $\dot{V}_0$ satisfies

$$\dot{V}_0 \leq \sum_{i=2}^{N} \eta_i^T \left(A^T P + PA - 2(\alpha - 1) PDD^T P + \kappa PP\right) \eta_i + \frac{1}{\kappa}\|\Psi\|^2, \tag{5.35}$$

with κ being any positive real number.

Using Lemmas 5.2.1 and 5.2.2, we easily obtain

$$\dot{V}_0 \leq \sum_{i=2}^{N} \eta_i^T \left(A^T P + PA - 2\alpha PDD^T P + \kappa PP + \frac{\gamma_0^2}{\kappa} I_n\right) \eta_i + \frac{4\gamma_0^2}{\kappa}\lambda_\sigma^2(Q)\Delta, \tag{5.36}$$

for Case (1) with $\Delta = \delta^T \delta$, and

$$\dot{V}_0 \leq \sum_{i=2}^{N} \eta_i^T \left(A^T P + PA - 2(\alpha - 1) PDD^T P + \kappa PP + \frac{\gamma_0^2}{\kappa} I_n\right) \eta_i + \frac{4\gamma_0^2}{\kappa}\lambda_\sigma^2(Q)\Delta, \tag{5.37}$$

for Case (2). Here we have used $\|\eta\|^2 = \sum_{i=2}^{N} \|\eta_i\|^2$.

The remaining analysis is to explore the bound of Δ. With δ_l in (5.25) and Lemma 2.1.2 in Chapter 2, we have

$$\begin{aligned}\Delta_i &= \int_t^{t+h} \eta_i^T(\tau-h)K^TB^T\mathrm{e}^{A^T(t-\tau)}\mathrm{d}\tau \int_t^{t+h} \mathrm{e}^{A(t-\tau)}BK\eta_i(\tau-h)\mathrm{d}\tau \\ &\leq h\int_t^{t+h} \eta_i^T(\tau-h)PDD^T\mathrm{e}^{A^Th}\mathrm{e}^{A^T(t-\tau)}\mathrm{e}^{A(t-\tau)}\mathrm{e}^{Ah}DD^TP\eta_i(\tau-h)\mathrm{d}\tau.\end{aligned}$$

In view of Lemma 2.1.1 in Chapter 2 with $P = I_n$, provided that

$$R = -A^T - A + \omega_1 I_n > 0, \tag{5.38}$$

we have

$$\mathrm{e}^{A^Tt}\mathrm{e}^{At} < \mathrm{e}^{\omega_1 t}I_n,$$

and

$$\begin{aligned}\Delta_i &\leq h\int_t^{t+h} \mathrm{e}^{\omega_1(t-\tau)}\eta_i^T(\tau-h)PDD^T\mathrm{e}^{A^Th}\mathrm{e}^{Ah}DD^TP\eta_i(\tau-h)\mathrm{d}\tau \\ &\leq \rho^2h\mathrm{e}^{\omega_1 h}\int_t^{t+h} \mathrm{e}^{\omega_1(t-\tau)}\eta_i^T(\tau-h)\eta_i(\tau-h)\mathrm{d}\tau \\ &\leq \rho^2h\mathrm{e}^{2\omega_1 h}\int_t^{t+h} \eta_i^T(\tau-h)\eta_i(\tau-h)\mathrm{d}\tau,\end{aligned}$$

where ρ is a positive real number satisfying

$$\rho^2 I_n \geq PDD^TDD^TP. \tag{5.39}$$

Then the summation of Δ_i can be obtained as

$$\Delta = \sum_{i=2}^{N}\Delta_i \leq \sum_{i=2}^{N}\rho^2h\mathrm{e}^{2\omega_1 h}\int_t^{t+h} \eta_i^T(\tau-h)\eta_i(\tau-h)\mathrm{d}\tau. \tag{5.40}$$

For the integral term Δ shown in (5.40), we consider the following Krasovskii functional

$$W_i = \int_t^{t+h} \mathrm{e}^{\tau-t}\eta_i^T(\tau-h)\eta_i(\tau-h)\mathrm{d}\tau + \int_t^{t+h} \eta_i^T(\tau-2h)\eta_i(\tau-2h)\mathrm{d}\tau.$$

A direct evaluation gives that

$$\begin{aligned}\dot{W}_i &= -\int_t^{t+h} \mathrm{e}^{\tau-t}\eta_i^T(\tau-h)\eta_i(\tau-h)\mathrm{d}\tau \\ &\quad - \eta_i(t-2h)^T\eta_i(t-2h) + \mathrm{e}^h\eta_i^T(t)\eta_i(t) \\ &\leq -\int_t^{t+h} \eta_i^T(\tau-h)\eta_i(\tau-h)\mathrm{d}\tau + \mathrm{e}^h\eta_i^T(t)\eta_i(t).\end{aligned}$$

With $W_0 = \sum_{i=2}^{N} W_i$, we have

$$\begin{aligned} \dot{W}_0 =& \sum_{i=2}^{N} \dot{W}_i \\ \le& - \sum_{i=2}^{N} \int_{t}^{t+h} \eta_i^T(\tau - h)\eta_i(\tau - h)\mathrm{d}\tau + \sum_{i=2}^{N} \mathrm{e}^h \eta_i^T(t)\eta_i(t). \end{aligned} \tag{5.41}$$

Let

$$V = V_0 + \rho^2 h \mathrm{e}^{2\omega_1 h} \frac{4\gamma_0^2}{\kappa} \lambda_\sigma^2(Q) W_0. \tag{5.42}$$

From (5.36), (5.37), (5.40) and (5.41), we obtain that

$$\dot{V} \le \eta^T(t) \left(I_N \otimes H\right) \eta(t), \tag{5.43}$$

where

$$H := A^T P + PA - 2\alpha PDD^T P + \kappa PP + \frac{\gamma_0^2}{\kappa} \left(1 + \lambda_\sigma^2(Q)\rho^2 h \mathrm{e}^{(2\omega_1 + 1)h}\right) I_n, \tag{5.44}$$

for Case (1), and

$$H := A^T P + PA - 2(\alpha - 1) PDD^T P + \kappa PP + \frac{\gamma_0^2}{\kappa} \left(1 + \lambda_\sigma^2(Q)\rho^2 h \mathrm{e}^{(2\omega_1 + 1)h}\right) I_n, \tag{5.45}$$

for Case (2).

The above expressions can be used for consensus analysis of network-connected systems with Lipschitz nonlinearity and input delay. The following theorem summarizes the results.

Theorem 9 *For an input-delayed multi-agent system (5.1) with the associated Laplacian matrix that satisfies Assumption 5.1.1, the consensus control problem can be solved by the control algorithm (5.7) with the control gain* $K = D^T P$ *specified in one of the following two cases:*

(1) If the eigenvalues of the Laplacian matrix $\mathcal{L}$ *are distinct, the consensus is achieved if the following conditions are satisfied for* $W = P^{-1}$ *and* $\rho > 0$, $\omega_1 \ge 0$,

$$(A - \frac{1}{2}\omega_1 I_n)^T + (A - \frac{1}{2}\omega_1 I_n) < 0, \tag{5.46}$$

$$\rho W \ge DD^T, \tag{5.47}$$

$$\begin{bmatrix} WA^T + AW - 2\alpha DD^T + \kappa I_n & W \\ W & \dfrac{-\kappa I_n}{\gamma_0^2(1 + 4h_0\rho^2)} \end{bmatrix} < 0, \tag{5.48}$$

where κ *is any positive real number and* $h_0 = \lambda_\sigma^2(Q)he^{(2\omega_1+1)h}$.

(2) If the Laplacian matrix $\mathcal{L}$ *has multiple eigenvalues, the consensus is achieved if the conditions (5.46), (5.47) and the following condition are satisfied for* $W = P^{-1}$ *and* $\rho > 0$, $\omega_1 \geq 0$,

$$\begin{bmatrix} WA^T + AW - 2(\alpha-1)DD^T + \kappa I_n & W \\ W & \frac{-\kappa I_n}{\gamma_0^2(1+4h_0\rho^2)} \end{bmatrix} < 0, \tag{5.49}$$

where κ *is any positive real number and* $h_0 = \lambda_\sigma^2(Q)he^{(2\omega_1+1)h}$.

Proof 15 *When the eigenvalues are distinct, from the analysis in this section, we know that the feedback law (5.7) will stabilize* η *if the conditions (5.38), (5.39) and* $H < 0$ *in (5.44) are satisfied. Indeed, it is easy to see the conditions (5.38) and (5.39) are equivalent to the conditions specified in (5.46) and (5.47). From (5.44), it can be obtained that* $H < 0$ *is equivalent to*

$$P^{-1}A^T + AP^{-1} - 2\alpha DD^T + \kappa I_n + \frac{\gamma_0^2}{\kappa}(1+4h_0\rho^2)P^{-1}P^{-1} < 0, \tag{5.50}$$

which is further equivalent to (5.48). Hence we conclude that η *converges to zero asymptotically.*

When the Laplacian matrix has multiple eigenvalues, the feedback law (5.7) will stabilize η *if the conditions (5.38), (5.39) and* $H < 0$ *in (5.45) are satisfied. Following the similar procedure as Case (1), we can show that, under the conditions (5.46), (5.47) and (5.49),* η *converges to zero asymptotically. The proof is completed.*

5.3 Truncated-Predictor-Based Consensus

Assumption 5.3.1 *The eigenvalues of the Laplacian matrix are distinct.*

Remark 5.3.1 *This assumption ensures that the eigenvalue zero is algebraically simple and the directed graph contains a spanning tree. A stronger condition imposed by Assumption 5.3.1 is for the convenience of presentation. The results shown in this section can be extended to the cases in last section where the Laplacian matrix has multiple eigenvalues other than at zero, by following the procedures shown in [26].*

For the multi-agent systems (5.1), we have

$$x_i(t) = e^{Ah}x_i(t-h) + \int_{t-h}^{t} e^{A(t-\tau)}\left(Bu_i(\tau-h) + \phi\left(x_i(\tau)\right)\right)d\tau.$$

We propose a control design based on the truncated prediction method. The control input takes the structure

$$u_i(t) = Ke^{Ah}\sum_{j=1}^{N} q_{ij}\left(x_i(t) - x_j(t)\right) = Ke^{Ah}\sum_{j=1}^{N} l_{ij}x_j(t), \tag{5.51}$$

where $K \in \mathbb{R}^{m\times n}$ is a constant control gain matrix to be designed later. Under control algorithm (5.51), the multi-agent systems (5.1) can be written as

$$\begin{aligned}\dot{x}_i(t) = {} & Ax_i(t) + BK\sum_{j=1}^{N} l_{ij}x_j + \phi(x_i) \\ & - BK\sum_{j=1}^{N} l_{ij}\int_{t-h}^{t} e^{A(t-\tau)}\left(Bu_j(\tau - h) + \phi(x_j)\right)\mathrm{d}\tau.\end{aligned}$$

The closed-loop system is then described by

$$\dot{x} = \left(I_N \otimes A + L \otimes BK\right)x + \left(L \otimes BK\right)\left(d_1 + d_2\right) + \Phi(x), \tag{5.52}$$

where

$$\begin{aligned} d_1 &= -\int_{t-h}^{t} e^{A(t-\tau)}Bu(\tau - h)\mathrm{d}\tau, \\ d_2 &= -\int_{t-h}^{t} e^{A(t-\tau)}\Phi(x)\mathrm{d}\tau, \end{aligned}$$

with

$$\begin{aligned} x(t) &= \left[x_1^T(t), x_2^T(t), \cdots, x_N^T(t)\right]^T, \\ u(t) &= \left[u_1^T(t), u_2^T(t), \cdots, u_N^T(t)\right]^T, \\ \Phi(x) &= \left[\phi^T(x_1), \phi^T(x_2), \cdots, \phi^T(x_N)\right]^T. \end{aligned}$$

Define $r^T \in \mathbb{R}^{1\times N}$ as the left eigenvector of the Laplacian matrix $\mathcal{L}$ associated with the eigenvalue zero, satisfying $r^T\mathcal{L} = 0$. Furthermore, let r be scaled such that $r^T\mathbf{1} = 1$. By Assumption 5.3.1, we know that a non-singular matrix T, with its first column $T_{(1)} = \mathbf{1}$ and the first row of T^{-1}, $T_{(1)}^{-1} = r^T$, can be constructed such that

$$T^{-1}\mathcal{L}T = J, \tag{5.53}$$

where J being a block diagonal matrix in the real Jordan form with the structure

$$J = \begin{bmatrix} 0 & & & & & & \\ & \lambda_1 & & & & & \\ & & \ddots & & & & \\ & & & \lambda_{n_\lambda} & & & \\ & & & & \nu_1 & & \\ & & & & & \ddots & \\ & & & & & & \nu_{n_\nu} \end{bmatrix},$$

where $\lambda_i \in \mathbb{R}$ for $i = 1, 2, \cdots, n_\lambda$, and

$$\nu_i = \begin{bmatrix} \alpha_i & \beta_i \\ -\beta_i & \alpha_i \end{bmatrix} \in \mathbb{R}^{2\times 2},$$

for $i = 1, 2, \ldots, n_\nu$. In the above expression of J, λ_i, α_i and β_i are positive real numbers with λ_i denotes real eigenvalues of $\mathcal{L}$, and $\alpha_i \pm i\beta_i$ represent conjugate complex eigenvalues of $\mathcal{L}$. Clearly we have $1 + n_\lambda + 2n_\nu = N$. Moreover, all the non-zero eigenvalues of $\mathcal{L}$ are positive or with positive real parts.

Based on the vector r, we introduce a state transformation

$$\xi_i = x_i - \sum_{j=1}^{N} r_j x_j, \tag{5.54}$$

for $i = 1, 2, \ldots, N$. Let $\xi = \left[\xi_1^T, \xi_2^T, \cdots, \xi_N^T\right]^T$. Then we have

$$\begin{aligned} \xi &= x - \left(\left(\mathbf{1}r^T\right) \otimes I_n\right) x \\ &= \left(M \otimes I_n\right) x, \end{aligned}$$

where $M = I_N - \mathbf{1}r^T$. Since $r^T\mathbf{1} = 1$, it can be shown that $M\mathbf{1} = 0$. Therefore the consensus of system (5.52) is achieved when $\lim_{t\to\infty} \xi(t) = 0$, as $\xi = 0$ implies that $x_1 = x_2 = \cdots = x_N$, due to the fact that the null space of M is span$\{\mathbf{1}\}$. The dynamics of ξ can then be derived as

$$\begin{aligned} \dot{\xi} =& \left(I_N \otimes A + \mathcal{L} \otimes BK\right) x - \left(\mathbf{1}r^T \otimes I_n\right) \left[I_N \otimes A + \mathcal{L} \otimes BK\right] x \\ &+ \left(M \otimes I_n\right) \left(\mathcal{L} \otimes BK\right) \left(d_1 + d_2\right) + \left(M \otimes I_n\right) \Phi(x) \\ =& \left(I_N \otimes A + \mathcal{L} \otimes BK\right) \xi + \left(M \otimes I_n\right) \Phi(x) + \left(\mathcal{L} \otimes BK\right) \left(d_1 + d_2\right), \end{aligned} \tag{5.55}$$

where we have used $r^T\mathcal{L} = 0$.

To explore the structure of $\mathcal{L}$, we propose another state transformation

$$\eta = \left(T^{-1} \otimes I_n\right) \xi, \tag{5.56}$$

with $\eta = \left[\eta_1^T, \eta_2^T, \cdots, \eta_N^T\right]^T$. Then we have

$$\dot{\eta} = \left(I_N \otimes A + J \otimes BK\right) \eta + \Pi(x) + \Delta(x) + \Psi(x), \tag{5.57}$$

where

$$\begin{aligned} \Pi(x) &= \left(T^{-1}\mathcal{L} \otimes BK\right) d_1, \\ \Delta(x) &= \left(T^{-1}\mathcal{L} \otimes BK\right) d_2, \\ \Psi(x) &= \left(T^{-1}M \otimes I_n\right) \Phi(x). \end{aligned}$$

For the notational convenience, let

$$\Pi = \begin{bmatrix} \pi_1 \\ \pi_2 \\ \vdots \\ \pi_N \end{bmatrix}, \quad \Delta = \begin{bmatrix} \delta_1 \\ \delta_2 \\ \vdots \\ \delta_N \end{bmatrix}, \quad \Psi = \begin{bmatrix} \psi_1 \\ \psi_2 \\ \vdots \\ \psi_N \end{bmatrix},$$

with $\pi_i, \delta_i, \psi_i, : \mathbb{R}^{n\times N} \to \mathbb{R}^n$ for $i = 1, 2, \ldots, N$.

From state transformations (5.54) and (5.56), we have:

$$\eta_1 = \left(r^T \otimes I_n\right)\xi = \left(\left(r^T M\right)\otimes I_n\right)x \equiv 0.$$

With the control law shown in (5.51), the control gain matrix K is chosen as

$$K = -B^T P, \tag{5.58}$$

where P is a positive definite matrix.

The following theorem presents sufficient conditions to ensure that the consensus problem is solved by using the control algorithm (5.51) with control gain K in (5.58).

Theorem 10 *For the Lipschitz nonlinear multi-agent systems (5.1) with input delay, the consensus control problem can be solved by the control algorithm (5.51) with $K = -B^T P$ if there exists a positive definite matrix P and constants $\omega_1 \geq 0$, $\rho, \kappa_1, \kappa_2, \kappa_3 > 0$ such that*

$$\rho W \geq BB^T, \tag{5.59}$$

$$\left(A - \frac{1}{2}\omega_1 I_n\right)^T + \left(A - \frac{1}{2}\omega_1 I_n\right) < 0, \tag{5.60}$$

$$\begin{bmatrix} WA^T + AW - 2\alpha BB^T + (\kappa_1 + \kappa_2 + \kappa_3)I_n & W \\ W & -\frac{I_n}{\Gamma} \end{bmatrix} < 0, \tag{5.61}$$

are satisfied with $W = P^{-1}$ and

$$\begin{aligned}
\alpha &= \min\{\lambda_1, \lambda_2, \cdots, \lambda_{n_\lambda}, \alpha_1, \alpha_2, \cdots, \alpha_{n_\nu}\}, \\
\Gamma &= \frac{\gamma_0}{\kappa_1}e^h + \frac{\gamma_1}{\kappa_2}e^h + \frac{\gamma_2}{\kappa_3}, \\
\gamma_0 &= 4h\rho^4 e^{2\omega_1 h}\lambda_\sigma^2\left(T^{-1}\right)\|\mathcal{L}\|_F^2\|Q\|_F^2\|T\|_F^2, \\
\gamma_1 &= 4\rho^2 h e^{\omega_1 h}\gamma^2\lambda_\sigma^2\left(T^{-1}\right)\lambda_\sigma^2(Q)\|T\|_F^2, \\
\gamma_2 &= 4N\gamma^2\|r\|^2\lambda_\sigma^2\left(T^{-1}\right)\|T\|_F^2,
\end{aligned}$$

where Q is the adjacency matrix, $\mathcal{L}$ is the Laplacian matrix, T is the non-singular matrix defined in (5.53), ρ and ω_1 are positive numbers such that

$$\rho^2 I \geq PBB^T BB^T P, \tag{5.62}$$

$$\omega_1 I > A^T + A, \tag{5.63}$$

and $\lambda_\sigma(\cdot)$ and $\|\cdot\|_F$ are the maximum singular value and the Frobenius norm of a matrix, respectively.

Proof 16 *The consensus analysis will be carried out in terms of η. By (5.56), the consensus is achieved if η converges to zero, or equivalently if η_i converges to zero for $i = 2, 3, \ldots, N$, since it has been shown that $\eta_1 = 0$. Let*

$$V_i = \eta_i^T P \eta_i, \tag{5.64}$$

for $i = 2, 3, \cdots, N$. For $i = 2, 3, \cdots, n_\lambda + 1$, we have

$$\dot{\eta}_i = \left(A - \lambda_i B B^T P\right) \eta_i + \pi_i + \delta_i + \psi_i,$$

and hence

$$\begin{aligned}
\dot{V}_i &= \eta_i^T \left(A^T P + PA - 2\lambda_i PBB^T P\right) \eta_i + 2\eta_i^T P \pi_i + 2\eta_i^T P \delta_i + 2\eta_i^T P \psi_i \\
&\leq \eta_i^T \left(A^T P + PA - 2\lambda_i PBB^T P + \sum_{\iota=1}^{3} \kappa_\iota PP\right) \eta_i \\
&\quad + \frac{1}{\kappa_1}\|\pi_i\|^2 + \frac{1}{\kappa_2}\|\delta_i\|^2 + \frac{1}{\kappa_3}\|\psi_i\|^2,
\end{aligned} \tag{5.65}$$

where κ_1, κ_2 and κ_3 are any positive numbers. In the above derivation, we used the well-known inequality

$$2a^T b \leq \kappa a^T a + \frac{1}{\kappa} b^T b, \quad \forall a, b \in \mathbb{R}^n.$$

For $i = n_\lambda + 2, n_\lambda + 3, \cdots, N$, we consider the evolution of the states in pairs corresponding to each pair of conjugate eigenvalues. For a $k \in \{1, 2, \cdots, n_\nu\}$, let

$$\begin{aligned}
i_1 &= 1 + n_\lambda + 2k - 1, \\
i_2 &= 1 + n_\lambda + 2k.
\end{aligned}$$

The dynamics of η_{i_1} and η_{i_2} are expressed as

$$\begin{aligned}
\dot{\eta}_{i_1} &= \left(A - \alpha_k BB^T P\right) \eta_{i_1} - \beta_k BB^T P \eta_{i_2} + \pi_{i_1} + \delta_{i_1} + \psi_{i_1}, \\
\dot{\eta}_{i_2} &= \left(A - \alpha_k BB^T P\right) \eta_{i_2} + \beta_k BB^T P \eta_{i_1} + \pi_{i_2} + \delta_{i_2} + \psi_{i_2}.
\end{aligned}$$

Let

$$V_i = \eta_{i_1}^T P \eta_{i_1} + \eta_{i_2}^T P \eta_{i_2}.$$

Using the dynamics shown above, we can compute in a similar way as in the real eigenvalue case that

$$\begin{aligned}\dot{V}_i &= \eta_{i_1}^T\left(A^TP+PA-2\alpha_k PBB^TP\right)\eta_{i_1}+\eta_{i_2}^T\left(A^TP+PA-2\alpha_k PBB^TP\right)\eta_{i_2}\\&\quad+2\eta_{i_1}^TP\pi_{i_1}+2\eta_{i_1}^TP\delta_{i_1}+2\eta_{i_1}^TP\psi_{i_1}+2\eta_{i_2}^TP\pi_{i_2}+2\eta_{i_2}^TP\delta_{i_2}+2\eta_{i_2}^TP\psi_{i_2}\\&\leq \eta_{i_1}^T\left(A^TP+PA-2\alpha_k PBB^TP+\sum_{\iota=1}^{3}\kappa_\iota PP\right)\eta_{i_1}\\&\quad+\frac{1}{\kappa_1}\|\pi_{i_1}\|^2+\frac{1}{\kappa_2}\|\delta_{i_1}\|^2+\frac{1}{\kappa_3}\|\psi_{i_1}\|^2\\&\quad+\eta_{i_2}^T\left(A^TP+PA-2\alpha_k PBB^TP+\sum_{\iota=1}^{3}\kappa_\iota PP\right)\eta_{i_2}\\&\quad+\frac{1}{\kappa_1}\|\pi_{i_2}\|^2+\frac{1}{\kappa_2}\|\delta_{i_2}\|^2+\frac{1}{\kappa_3}\|\psi_{i_2}\|^2,\end{aligned}\tag{5.66}$$

where κ_1,κ_2 and κ_3 are any positive numbers. Let

$$V_0=\sum_{i=2}^{N}V_i.$$

In view of (5.65) and (5.66), we have

$$\begin{aligned}\dot{V}_0 &= \sum_{i=2}^{n_\lambda+1}\dot{V}_i+\sum_{i=n_\lambda+2}^{N}\dot{V}_i\\&\leq \eta^T\left[I_N\otimes\left(A^TP+PA-2\alpha PBB^TP+\sum_{\iota=1}^{3}\kappa_\iota PP\right)\right]\eta\\&\quad+\frac{1}{\kappa_1}\|\Pi\|^2+\frac{1}{\kappa_2}\|\Delta\|^2+\frac{1}{\kappa_3}\|\Psi\|^2.\end{aligned}\tag{5.67}$$

Lemma 5.3.1 *For the integral terms $\|\Pi\|^2$ and $\|\Delta\|^2$ shown in the transformed system dynamics (5.57), the bounds can be established as*

$$\|\Pi\|^2\leq\gamma_0\int_{t-h}^{t}\eta^T(\tau-h)\,\eta(\tau-h)\,\mathrm{d}\tau,\tag{5.68}$$

$$\|\Delta\|^2\leq\gamma_1\int_{t-h}^{t}\eta^T(\tau)\eta(\tau)\mathrm{d}\tau,\tag{5.69}$$

if the conditions (5.62) and (5.63) are satisfied.

Proof 17 *By the definition of $\Pi(x)$ in (5.57), we have*

$$\|\Pi\|=\left\|\left(T^{-1}\otimes I_n\right)(\mathcal{L}\otimes BK)d_1\right\|\leq\lambda_\sigma\left(T^{-1}\right)\|\mu\|,\tag{5.70}$$

where $\mu = (\mathcal{L}\otimes BK)d_1$. For the notational convenience, let $\mu =$

$\left[\mu_1^T, \mu_2^T, \ldots, \mu_N^T\right]^T$. *Then from (5.52) and (5.58), we have*

$$\begin{aligned}\mu_i =& -BK\sum_{j=1}^{N} l_{ij}\int_{t-h}^{t} e^{A(t-\tau)}Bu_j(\tau-h)\mathrm{d}\tau \\ =& BB^TP\sum_{j=1}^{N} l_{ij}\int_{t-h}^{t} e^{A(t-\tau)}BB^TPe^{Ah}\sum_{k=1}^{N} q_{jk}\left(x_k(\tau-h)-x_j(\tau-h)\right)\mathrm{d}\tau.\end{aligned} \tag{5.71}$$

From $\eta = \left(T^{-1}\otimes I_n\right)\xi$, *we obtain* $\xi = \left(T\otimes I_n\right)\eta$, *and from the state transformation (5.54), we have*

$$\begin{aligned}x_k(t)-x_j(t) &= \xi_k(t)-\xi_j(t) \\ &= \left((T_k-T_j)\otimes I_n\right)\eta(t) \\ &= \sum_{l=1}^{N}\left(T_{kl}-T_{jl}\right)\eta_l(t),\end{aligned} \tag{5.72}$$

where T_k *denotes the* k*th row of* T. *We define*

$$\sigma_l = BB^TP\int_{t-h}^{t} e^{A(t-\tau)}BB^TPe^{Ah}\eta_l(\tau-h)\mathrm{d}\tau.$$

Then, from (5.71) and (5.72), we can obtain that

$$\mu_i = \sum_{j=1}^{N} l_{ij}\sum_{k=1}^{N} q_{jk}\sum_{l=1}^{N}\left(T_{kl}-T_{jl}\right)\sigma_l.$$

For the notational convenience, let $\sigma = \left[\sigma_1^T, \sigma_2^T, \ldots, \sigma_N^T\right]^T$. *It then follows that*

$$\begin{aligned}\|\mu_i\| \leq& \sum_{j=1}^{N}|l_{ij}|\sum_{k=1}^{N}|q_{jk}|\,\|T_k\|\,\|\sigma\| + \sum_{k=1}^{N}\sum_{j=1}^{N}|l_{ij}|\,|q_{jk}|\,\|T_j\|\,\|\sigma\| \\ \leq& \sum_{j=1}^{N}|l_{ij}|\,\|q_j\|\,\|T\|_F\,\|\sigma\| + \sum_{k=1}^{N}\sum_{j=1}^{N}|l_{ij}|\,\|q_k\|\,\|T\|_F\,\|\sigma\| \\ \leq& \|l_i\|\,\|Q\|_F\,\|T\|_F\,\|\sigma\| + \|l_i\|\,\|Q\|_F\,\|T\|_F\,\|\sigma\| \\ =& 2\|l_i\|\,\|Q\|_F\,\|T\|_F\,\|\sigma\|,\end{aligned}$$

where l_i *denotes the* i*th row of* L. *Therefore we have*

$$\begin{aligned}\|\mu\|^2 &= \sum_{i=1}^{N}\|\mu_i\|^2 \\ &\leq 4\sum_{i=1}^{N}\|l_i\|^2\,\|Q\|_F^2\,\|T\|_F^2\,\|\sigma\|^2 \\ &= 4\|\mathcal{L}\|_F^2\,\|Q\|_F^2\,\|T\|_F^2\,\|\sigma\|^2.\end{aligned} \tag{5.73}$$

Now we need to deal with $\|\sigma\|^2$. *With Lemma 2.1.2 and the condition (5.62), we have*

$$\begin{aligned}\|\sigma_i\|^2 \le\, & h \int_{t-h}^{t} \eta_i^T(\tau-h)e^{A^T h}PBB^T e^{A^T(t-\tau)}PBB^T \\ & \times BB^T Pe^{A(t-\tau)}BB^T Pe^{Ah}\eta_i(\tau-h)\mathrm{d}\tau \\ \le\, & h\rho^2 \int_{t-h}^{t} \eta_i^T(\tau-h)e^{A^T h}PBB^T e^{A^T(t-\tau)}e^{A(t-\tau)}BB^T Pe^{Ah}\eta_i(\tau-h)\mathrm{d}\tau,\end{aligned}$$

In view of Lemma 2.1.1, with the condition (5.63), we have

$$\begin{aligned}\|\sigma_i\|^2 \le\, & h\rho^2 \int_{t-h}^{t} e^{\omega_1(t-\tau)}\eta_i^T(\tau-h)e^{A^T h}PBB^T BB^T Pe^{Ah}\eta_i(\tau-h)\mathrm{d}\tau \\ \le\, & h\rho^4 e^{\omega_1 h}\int_{t-h}^{t} \eta_i^T(\tau-h)e^{A^T h}e^{Ah}\eta_i(\tau-h)\mathrm{d}\tau \\ \le\, & h\rho^4 e^{2\omega_1 h}\int_{t-h}^{t} \eta_i^T(\tau-h)\eta_i(\tau-h)\mathrm{d}\tau.\end{aligned}$$

Then, $\|\sigma\|^2$ *can be bounded as*

$$\|\sigma\|^2 = \sum_{i=1}^{N}\|\sigma_i\|^2 \le\; h\rho^4 e^{2\omega_1 h}\int_{t-h}^{t} \eta^T(\tau-h)\eta(\tau-h)\mathrm{d}\tau. \tag{5.74}$$

Therefore, from (5.70), (5.73) and (5.74), we have

$$\|\Pi\|^2 \le \gamma_0 \int_{t-h}^{t} \eta^T(\tau-h)\eta(\tau-h)\mathrm{d}\tau.$$

This completes the proof. The proof of $\|\Delta\|$ *is similar to that of* $\|\Pi\|$ *and hence omitted.*

Lemma 5.3.2 *For the nonlinear term* $\Psi(x)$ *in the transformed system dynamics (5.57), a bound can be established as*

$$\|\Psi\|^2 \le \gamma_2 \|\eta\|^2, \tag{5.75}$$

where γ_2 *is defined in Theorem 10.*

Proof 18 *By the definition of* $\Psi(x)$ *in (5.57), we have*

$$\|\Psi\| = \left\|\left(T^{-1}\otimes I_n\right)\left(M\otimes I_n\right)\Phi(x)\right\| \le \lambda_\sigma\left(T^{-1}\right)\|z\|,$$

where $z = (M\otimes I_n)\Phi(x)$. *For the notational convenience, let* $z = \left[z_1^T, z_2^T, \ldots, z_N^T\right]^T$. *Then from (5.54), we have*

$$z_i = \phi(x_i) - \sum_{k=1}^{N} r_k\phi(x_k) = \sum_{k=1}^{N} r_k(\phi(x_i)-\phi(x_k)).$$

It then follows that

$$\|z_i\| \leq \sum_{k=1}^{N} |r_k| \, \|(\phi(x_i) - \phi(x_k))\| \leq \gamma \sum_{k=1}^{N} |r_k| \, \|x_i - x_k\| .$$

In light of (5.72), we have

$$\|z_i\| \leq \gamma \sum_{k=1}^{N} |r_k| \left(\|T_i\| + \|T_k\|\right) \|\eta\| \leq \gamma \|\eta\| \left(\sum_{k=1}^{N} |r_k| \, \|T_i\| + \|r\| \, \|T\|_F \right).$$

Therefore we have

$$\begin{aligned}
\|z\|^2 &= \sum_{i=1}^{N} \|z_i\|^2 \\
&\leq 2\gamma^2 \|\eta\|^2 \sum_{i=1}^{N} \left(\|T_i\|^2 \left(\sum_{k=1}^{N} |r_k| \right)^2 + \|r\|^2 \|T\|_F^2 \right) \\
&\leq 2\gamma^2 \|\eta\|^2 \sum_{i=1}^{N} \left(\|T_i\|^2 N \|r\|^2 + \|r\|^2 \|T\|_F^2 \right) \\
&= 4N\gamma^2 \|r\|^2 \|T\|_F^2 \|\eta\|^2 ,
\end{aligned}$$

and

$$\|\Psi\|^2 \leq \gamma_2 \|\eta\|^2 .$$

This completes the proof.

Using (5.67), (5.68), (5.69) and (5.75), we can obtain

$$\begin{aligned}
\dot{V}_0 \leq &\eta^T \left[I_N \otimes \left(A^T P + PA - 2\alpha PBB^T P + \sum_{\iota=1}^{3} \kappa_\iota PP + \frac{\gamma_2}{\kappa_3} I_n \right) \right] \eta \\
&+ \frac{\gamma_0}{\kappa_1} \int_{t-h}^{t} \eta^T(\tau - h)\eta(\tau - h) \mathrm{d}\tau + \frac{\gamma_1}{\kappa_2} \int_{t-h}^{t} \eta^T(\tau)\eta(\tau) \mathrm{d}\tau. \qquad (5.76)
\end{aligned}$$

For the first integral term shown in (5.76), we consider the following Krasovskii functional

$$W_1 = e^h \int_{t-h}^{t} e^{\tau - t} \eta^T(\tau - h)\eta(\tau - h) \mathrm{d}\tau + e^h \int_{t-h}^{t} \eta^T(\tau)\eta(\tau) \mathrm{d}\tau.$$

A direct evaluation gives that

$$\begin{aligned}
\dot{W}_1 = &- e^h \int_{t-h}^{t} e^{\tau - t} \eta^T(\tau - h)\eta(\tau - h) \mathrm{d}\tau \\
&- \eta^T(t - 2h)\eta(t - 2h) + e^h \eta^T(t)\eta(t) \\
\leq &- \int_{t-h}^{t} \eta^T(\tau - h)\eta(\tau - h) \mathrm{d}\tau + e^h \eta(t)^T \eta(t). \qquad (5.77)
\end{aligned}$$

For the second integral term shown in (5.76), we consider the following Krasovskii functional

$$W_2 = e^h \int_{t-h}^{t} e^{\tau - t} \eta^T(\tau)\eta(\tau)\mathrm{d}\tau.$$

A direct evaluation gives that

$$\begin{aligned}\dot{W}_2 =& - e^h \int_{t-h}^{t} e^{\tau - t} \eta^T(\tau)\eta(\tau)\mathrm{d}\tau + e^h\eta^T(t)\eta(t) - \eta^T(t-h)\eta(t-h) \\ \le& - \int_{t-h}^{t} \eta^T(\tau)\eta(\tau)\mathrm{d}\tau + e^h\eta^T(t)\eta(t).\end{aligned} \tag{5.78}$$

Let

$$V = V_0 + \frac{\gamma_0}{\kappa_1}W_1 + \frac{\gamma_1}{\kappa_2}W_2. \tag{5.79}$$

From (5.76), (5.77) and (5.78), we obtain that

$$\dot{V} \le \eta^T(t)\left(I_N \otimes H\right)\eta(t), \tag{5.80}$$

where

$$H := A^T P + PA - 2\alpha PBB^T P + \sum_{\iota=1}^{3} \kappa_\iota PP + \left(\frac{\gamma_0}{\kappa_1}e^h + \frac{\gamma_1}{\kappa_2}e^h + \frac{\gamma_2}{\kappa_3}\right) I_n. \tag{5.81}$$

From the analysis in this section, we know that the feedback law (5.51) will stabilize η if the conditions (5.62), (5.63) and $H < 0$ in (5.80) are satisfied. Indeed, it is easy to see the conditions (5.62) and (5.63) are equivalent to the conditions specified in (5.59) and (5.60). From (5.81), it can be obtained that $H < 0$ is equivalent to

$$WA^T + AW - 2\alpha BB^T + (\kappa_1 + \kappa_2 + \kappa_3) I_n + \left(\frac{\gamma_0}{\kappa_1}e^h + \frac{\gamma_1}{\kappa_2}e^h + \frac{\gamma_2}{\kappa_3}\right) WW < 0,$$

which is further equivalent to (5.61). It implies that η converges to zero asymptotically. Hence, the consensus control is achieved.

It is observed that (5.61) is more likely to be satisfied if the values of $\rho, \omega_1, \kappa_1, \kappa_2$ and κ_3 are small. Therefore, the algorithm for finding a feasible solution of the conditions shown in (5.59) to (5.61) can be designed by following a step by step algorithm.

(1) Set $\omega_1 = \lambda_{\max}(A + A^T)$ if $\lambda_{\max}(A + A^T) > 0$; otherwise set $\omega_1 = 0$.

(2) Fix the value of $\rho, \omega_1, \kappa_1, \kappa_2$ and κ_3 to some constants $\tilde{\omega}_1 > \omega_1$ and $\tilde{\rho}, \tilde{\kappa}_1, \tilde{\kappa}_2, \tilde{\kappa}_3 > 0$; make an initial guess for the values of $\tilde{\rho}, \tilde{\omega}_1, \tilde{\kappa}_1, \tilde{\kappa}_2, \tilde{\kappa}_3$.

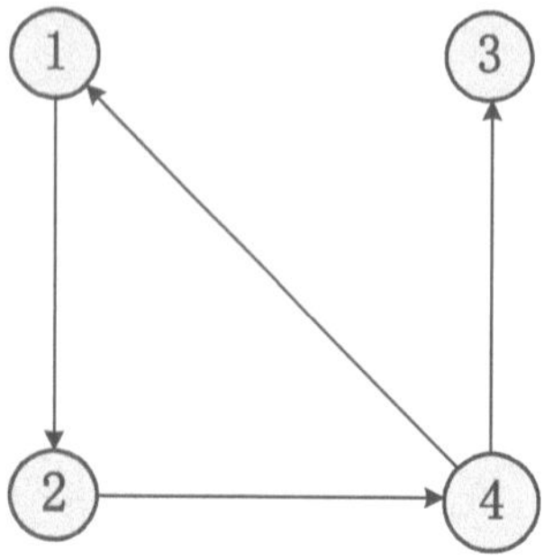

FIGURE 5.1: Communication topology.

(3) Solve the LMI equation (5.61) for W with the fixed values; if a feasible value of W cannot be found, return to Step (2) and reset the values of $\tilde{\rho}, \tilde{\omega}_1, \tilde{\kappa}_1, \tilde{\kappa}_2$ and $\tilde{\kappa}_3$.

(4) Solve the LMI equation (5.59) for ρ with the feasible value of W obtained in Step (3) and make sure that the value of ρ is minimized.

(5) If the condition $\tilde{\rho} \geq \rho$ is satisfied, then $(\tilde{\rho}, \tilde{\omega}_1, \tilde{\kappa}_1, \tilde{\kappa}_2, \tilde{\kappa}_3, W)$ is a feasible solution for Theorem 10; otherwise, set $\tilde{\rho} = \rho$ and return to Step (3).

Remark 5.3.2 *Given the input delay h and the Lipschitz constant γ, it is concluded that the existence of a feasible solution is related to the matrices (A, B) and the Laplacian matrix $\mathcal{L}$. Additionally, since the values of h and γ are fixed and they are not the decision variables of the LMIs, a feasible solution may not exist if the values of h and γ are too large. Therefore, a trigger should be added in the algorithm to stop the iteration procedure if the values of $\tilde{\rho}, \tilde{\omega}_1, \tilde{\kappa}_1, \tilde{\kappa}_2$* and and *$\tilde{\kappa}_3$ are out of the preset range.*

5.4 Numerical Examples

5.4.1 Predictor Case

In this section, we will illustrate in some details the proposed consensus control design through a circuit example. The system under consideration is a connection of four agents (i.e. $N = 4$) as shown in Figure 5.1, each of which is described by a second-order dynamic model as

$$\begin{cases} \dot{p}_i(t) = v_i(t), \\ \dot{v}_i(t) = f(v_i) + u_i(t-h), \end{cases} \tag{5.82}$$

where $p_i = [p_{ix}, p_{iy}, p_{iz}]^T \in \mathbb{R}^3$ denotes the position vector of agent i, $v_i = [v_{ix}, v_{iy}, v_{iz}]^T \in \mathbb{R}^3$ the velocity vector, $f(v_i) \in \mathbb{R}^3$ the intrinsic dynamics of agent i, governed by the chaotic Chua circuit [172]

$$f(v_i) = \begin{bmatrix} -0.59v_{ix} + v_{iy} - 0.17(|v_{ix}+1| - |v_{ix}-1|) \\ v_{ix} - v_{iy} + v_{iz} \\ -v_{iy} - 5v_{iz} \end{bmatrix}.$$

Let $x_i = [p_i^T, v_i^T]^T \in \mathbb{R}^6$. The dynamic equation (5.82) of each agent can be re-arranged as the state space model (5.2) with

$$A = \begin{bmatrix} 0 & 0 & 0 & 1 & 0 & 0 \\ 0 & 0 & 0 & 0 & 1 & 0 \\ 0 & 0 & 0 & 0 & 0 & 1 \\ 0 & 0 & 0 & -0.59 & 1 & 0 \\ 0 & 0 & 0 & 1 & -1 & 1 \\ 0 & 0 & 0 & 0 & -1 & -5 \end{bmatrix}, \quad B = \begin{bmatrix} 0 & 0 & 0 \\ 0 & 0 & 0 \\ 0 & 0 & 0 \\ 1 & 0 & 0 \\ 0 & 1 & 0 \\ 0 & 0 & 1 \end{bmatrix},$$

and $\phi(x_i) = [0, 0, 0, -0.17(|v_{ix}+1| - |v_{ix}-1|), 0, 0]^T$. The adjacency matrix is given by

$$Q = \begin{bmatrix} 0 & 0 & 0 & 1 \\ 1 & 0 & 0 & 0 \\ 0 & 0 & 0 & 1 \\ 0 & 1 & 0 & 0 \end{bmatrix},$$

and the resultant Laplacian matrix is obtained as

$$\mathcal{L} = \begin{bmatrix} 1 & 0 & 0 & -1 \\ -1 & 1 & 0 & 0 \\ 0 & 0 & 1 & -1 \\ 0 & -1 & 0 & 1 \end{bmatrix}.$$

The eigenvalues of $\mathcal{L}$ are $\{0, 1, 3/2 \pm j\sqrt{3}/2\}$, and therefore Assumption 5.1.1 is satisfied. Furthermore, the eigenvalues are distinct. We obtain that

$$J = \begin{bmatrix} 0 & 0 & 0 & 0 \\ 0 & 1 & 0 & 0 \\ 0 & 0 & \frac{3}{2} & \frac{\sqrt{3}}{2} \\ 0 & 0 & -\frac{\sqrt{3}}{2} & \frac{3}{2} \end{bmatrix},$$

with the matrices

$$T = \begin{bmatrix} 1 & 0 & \frac{1}{2} & \frac{\sqrt{3}}{2} \\ 1 & 0 & 1 & 0 \\ 1 & -2 & \frac{1}{2} & \frac{\sqrt{3}}{2} \\ 1 & 0 & \frac{1}{2} & -\frac{\sqrt{3}}{2} \end{bmatrix},$$

and $r^T = [1/3, 1/3, 0, 1/3]^T$.

The nonlinear function $\phi(x_i)$ in each agent dynamics is globally Lipschitz with a Lipschitz constant $\gamma = 0.34$, which gives $\gamma_0 = 3.7391$ by (5.15). Based on the Laplacian matrix $\mathcal{L}$, we have $\alpha = 1$. In simulation, the input delay is set as $h = 0.03$s. A positive definite matrix P can be obtained with $\kappa = 0.01$, $\omega_1 = 1.5$ and $\rho = 2$, as

$$P = \begin{bmatrix} 5.03 & -0.53 & 0.18 & 2.58 & 0.29 & 0.08 \\ -0.53 & 5.37 & 0.43 & 0.28 & 2.39 & 0.47 \\ 0.18 & 0.43 & 7.75 & -0.08 & -0.38 & 1.58 \\ 2.58 & 0.28 & -0.08 & 2.65 & 0.93 & 0.17 \\ 0.29 & 2.39 & -0.38 & 0.93 & 2.17 & 0.25 \\ 0.08 & 0.47 & 1.58 & 0.17 & 0.25 & 0.92 \end{bmatrix},$$

to satisfy the conditions of Theorem 9. Consequently, the control gain is obtained as

$$K = \begin{bmatrix} -2.19 & -0.12 & -0.01 & -2.46 & -0.74 & -0.15 \\ -0.13 & -2.10 & 0.30 & -0.75 & -2.08 & -0.32 \\ -0.09 & -0.43 & -1.64 & -0.18 & -0.18 & -1.27 \end{bmatrix}.$$

Simulation study has been carried out with the results shown in Figure 5.2 for the positions state disagreement of each agent. Clearly the conditions specified in Theorem 9 are sufficient for the control gain to achieve consensus control for the multi-agent systems. The same control gain has also been used for different values of input delay. The results shown in Figure 5.3 indicate that the conditions could be conservative in the control gain design for a given input delay and Lipschitz nonlinear function. Indeed, extensive simulation shows that the same control gain can possibly achieve consensus control for the system with a much larger delay and Lipschitz constant.

5.4.2 TPF Case

In this section, a simulation study is carried out to demonstrate the effectiveness of the proposed control design. Consider the same network connection as shown in Figure 5.1. The dynamics of the ith agent is described by a second-order model as

$$\dot{x}_i(t) = \begin{bmatrix} -0.09 & 1 \\ -1 & -0.09 \end{bmatrix} x_i(t) + g \begin{bmatrix} \sin(x_{i1}(t)) \\ 0 \end{bmatrix} + \begin{bmatrix} 0 \\ 1 \end{bmatrix} u(t-0.1).$$

The linear part of the system represents a decayed oscillator. The time delay of the system is 0.1 s, and the Lipschitz constant $\gamma = g$. The eigenvalues of L are $\{\ 0,\ 1,\ 3/2 \pm j\sqrt{3}/2\ \}$. Therefore, Assumption 5.3.1 is satisfied. We obtain that

$$J = \begin{bmatrix} 0 & 0 & 0 & 0 \\ 0 & 1 & 0 & 0 \\ 0 & 0 & \frac{3}{2} & \frac{\sqrt{3}}{2} \\ 0 & 0 & -\frac{\sqrt{3}}{2} & \frac{3}{2} \end{bmatrix},$$

with $\alpha = 1$ and $r^T = \left[\ \frac{1}{3},\ \frac{1}{3},\ 0,\ \frac{1}{3}\ \right]$.

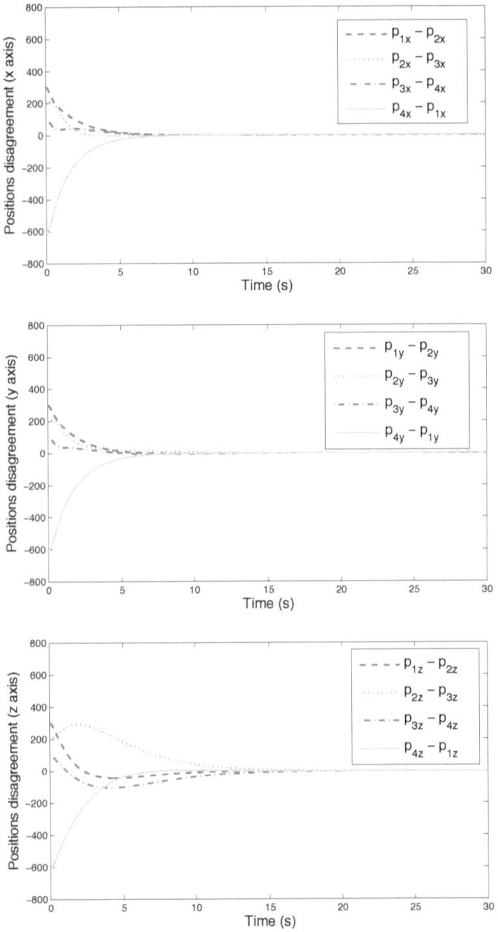

FIGURE 5.2: The positions disagreement of 4 agents: $h = 0.03\,\text{s}$.

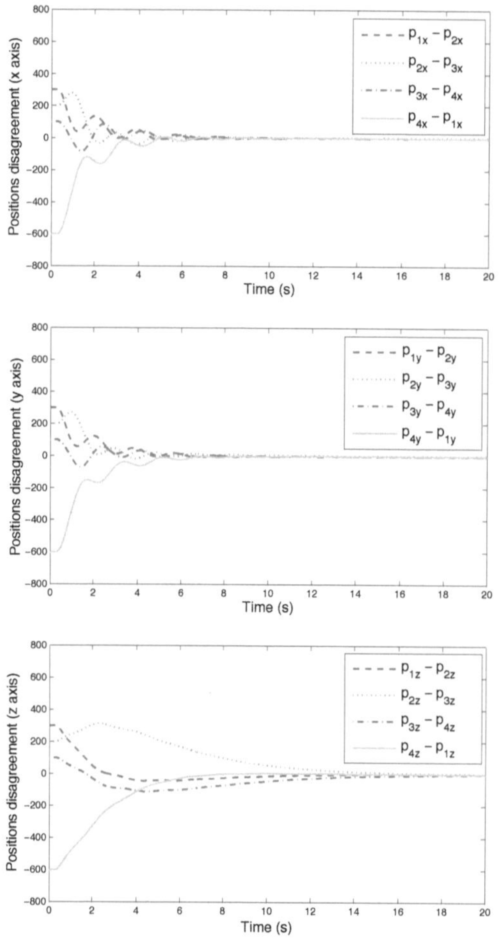

FIGURE 5.3: The positions disagreement of 4 agents: $h = 0.3\,\mathrm{s}$.

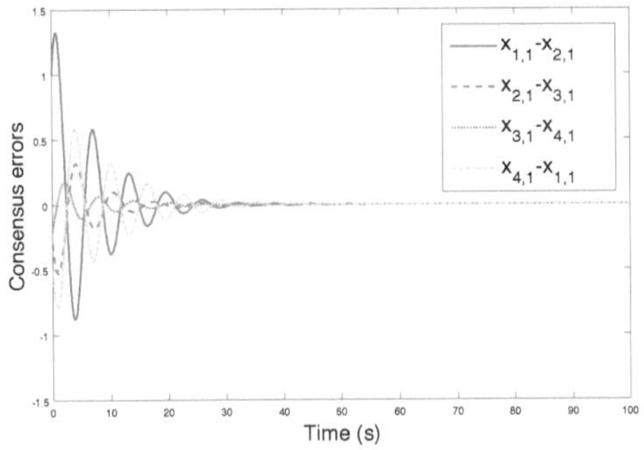

FIGURE 5.4: The consensus errors of state 1 with $h = 0.1$ and $g = 0.03$.

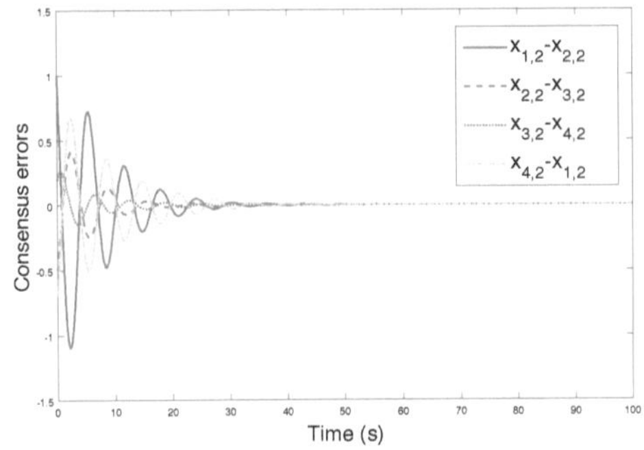

FIGURE 5.5: The consensus errors of state 2 with $h = 0.1$ and $g = 0.03$.

In this case, we choose $\gamma = g = 0.03$, and the initial conditions for the agents as $x_1(\theta) = [1,\ 1]^T$, $x_2(\theta) = [0,\ 0]^T$, $x_3(\theta) = [0.3,\ 0.5]^T$, $x_4(\theta) = [0.5,\ 0.3]^T, u(\theta) = [0,\ 0,\ 0,\ 0]^T$, for $\theta \in [-h, 0]$. With the values of $\omega_1 = 0, \rho = 0.05, \kappa_1 = \kappa_2 = 0.01$, and $\kappa_3 = 0.1$, a feasible solution of the feedback gain K is found to be

$$K = \left[\begin{array}{cc} -0.0021 & -0.0658 \end{array}\right].$$

Figures 5.4 and 5.5 show the simulation results for the consensus errors of all the agents. Clearly the conditions specified in Theorem 10 are sufficient for the control gain to achieve consensus control. Without re-tuning the control gain, the consensus control is still achieved for the multi-agent systems with a larger delay of 0.5 s and a bigger Lipschitz constant of $g = 0.15$, as shown in Figures 5.6 and 5.7, which imply the conditions might be conservative in the control gain design for a given input delay and Lipschitz condition.

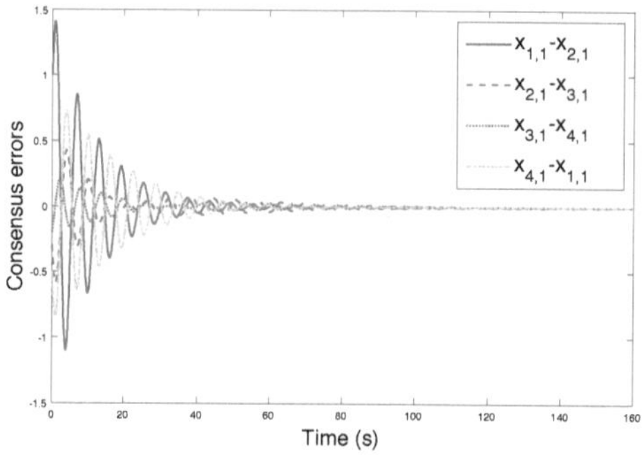

FIGURE 5.6: The consensus errors of state 1 with $h = 0.5$ and $g = 0.15$.

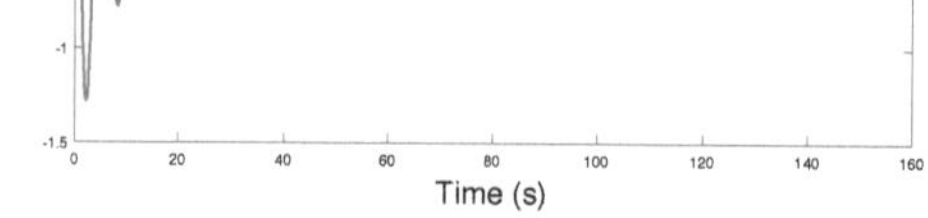

FIGURE 5.7: The consensus errors of state 2 with $h = 0.5$ and $g = 0.15$.

5.5 Conclusions

This chapter has investigated the impacts of nonlinearity and input delay in consensus control. This input delay may represent some delays in the network communication. We propose consensus protocols based on the predictor feedback and truncated prediction feedback for a class of Lipschitz nonlinear multi-agent systems with input delay. A complete consensus analysis is presented in a systematic framework of Lyapunov-Krasovskii functionals. Sufficient conditions are derived for the multi-agent systems to guarantee the global consensus in the time domain. The conditions can be solved by employing LMIs with a set of iterative parameters.

5.6 Notes

This chapter extends an idea in [26, 30] and develops consensus protocols for Lipschitz nonlinear multi-agent systems with input delay and directed topology. The main materials of this chapter are based on [161, 162]. It is worth mentioning that leaderless topology is considered in this chapter. The results in this chapter can be extended to leader-follower cases by following the procedures shown in [163, 164].

Chapter 6

Consensus Disturbance Rejection for Lipschitz Nonlinear MASs with Input Delay: A Predictor Feedback Approach

In this chapter, a predictor-based consensus disturbance rejection method is proposed for general multi-agent systems with Lipschitz nonlinearity and input delay. First, a distributed disturbance observer for consensus control is developed for each agent to estimate the disturbance under the delay constraint. Based on the conventional predictor feedback approach, a non-ideal predictor based control scheme is constructed for each agent by utilizing the estimate of the disturbance and the prediction of the relative state information. Then, rigorous analysis is carried out to ensure that the extra terms associated with disturbances and nonlinear functions are properly considered. Sufficient conditions for the consensus of the multi-agent systems with disturbance rejection are derived based on the analysis in the framework of Lyapunov-Krasovskii functionals. A simulation example is included to demonstrate the performance of the proposed control scheme.

6.1 Problem Formulation

In this chapter, we consider the leader-follower consensus control of a group of N agents. Assume that the dynamics of followers, labelled as $2, 3, \ldots, N$, are described by

$$\dot{x}_i(t) = Ax_i(t) + \phi(x_i(t)) + Bu_i(t-h) + D\omega_i(t), \tag{6.1}$$

and the leader agent is indexed by 1, whose dynamics is represented by

$$\dot{x}_1(t) = Ax_1(t) + \phi(x_1(t)), \tag{6.2}$$

where $x_i \in \mathbb{R}^n$ denotes the state, $u_i \in \mathbb{R}^m$ denotes the control input, $x_1 \in \mathbb{R}^n$ is the leader's state, $A \in \mathbb{R}^{n\times n}$ and $B \in \mathbb{R}^{n\times m}$ are constant matrices with

(A, B) being controllable, $h \in \mathbb{R}_+$ is the constant and known input delay, $\omega_i \in \mathbb{R}^s$ is a disturbance that is generated by a linear exogenous system

$$\dot{\omega}_i(t) = S\omega_i(t), \tag{6.3}$$

with $S \in \mathbb{R}^{s\times s}$ being a known constant matrix, and the nonlinear function $\phi : \mathbb{R}^n \to \mathbb{R}^n$, $\phi(0) = 0$, is assumed to satisfy the Lipschitz condition as

$$\|\phi(\alpha) - \phi(\beta)\| \leq \gamma\|\alpha - \beta\|, \forall\alpha, \beta \in \mathbb{R}^n$$

where $\gamma > 0$ is the Lipschitz constant.

Assumption 6.1.1 *The disturbance is matched. i.e., there exists a matrix $F \in \mathbb{R}^{m\times s}$ such that $D = BF$.*

Assumption 6.1.2 *The communication topology $\mathcal{G}$ contains a directed spanning tree with the leader as the root.*

Remark 6.1.1 *The matching condition in Assumption 6.1.1 guarantees that the disturbance act via the same channel as that of the control input. This assumption could be relaxed in some circumstances because unmatched disturbances under uncertain conditions may be converted to the matched ones based on output regulation theory [28]. Furthermore, the disturbance condition given in* (6.3) *is commonly used for disturbance rejection and output regulation. Many kinds of disturbances in engineering can be described by this model. For instance, unknown constant disturbances or harmonics with unknown amplitudes and phases, belong to the allowed class of disturbances.*

Because the leader has no neighbours, the Laplacian matrix $\mathcal{L}$ of $\mathcal{G}$ has the following structure

$$\mathcal{L} = \begin{bmatrix} 0 & 0_{1\times(N-1)} \\ \mathcal{L}_2 & \mathcal{L}_1 \end{bmatrix},$$

where $\mathcal{L}_1 \in \mathbb{R}^{(N-1)\times(N-1)}$ and $\mathcal{L}_2 \in \mathbb{R}^{(N-1)\times 1}$. From Lemma 1.4.8, it is obvious that $\mathcal{L}_1$ is a nonsingular M-matrix. We also have the following result for $\mathcal{L}_1$:

Lemma 6.1.1 ([71]) *For the nonsingular M-matrix $\mathcal{L}_1$, there exists a positive diagonal matrix G such that*

$$G\mathcal{L}_1 + \mathcal{L}_1^T G \geq r_0 I, \tag{6.4}$$

for some positive constant r_0. It is also shown that G can be constructed by letting $G = \text{diag}\{q_2, q_3, \cdots, q_N\} = (\text{diag}(\pi))^{-1}$, where $\pi = [\pi_2, \pi_3, \cdots, \pi_N]^T = (\mathcal{L}_1^T)^{-1}[1, 1, \cdots, 1]^T$.

Let $\xi_i = x_i - x_1, i = 2, 3, \cdots, N$ as the tracking errors. Then, based on the system dynamics (6.1) and (6.2), the error dynamics of the ith agent can be obtained as

$$\dot{\xi}_i(t) = A\xi_i(t) + \psi_i + Bu_i(t-h) + D\omega_i(t), \tag{6.5}$$

where $\psi_i = \phi(x_i(t)) - \phi(x_1(t))$.

With the agent 1 as the leader, the control objective of this chapter is to design a control input for each agent to follow the state of the leader x_1 under the disturbances. That is, under these control inputs, the following hold for any initial conditions,

$$\lim_{t\to\infty} (x_i(t) - x_1(t)) = \lim_{t\to\infty} \xi_i(t) = 0, \quad \forall i = 2, 3, \cdots, N.$$

6.2 Consensus Disturbance Rejection

The disturbance rejection design consists of disturbance estimation and rejection. The estimation is based on the relative state information obtained through the communication network. It is assumed that the ith agent collects the relative state information of its neighbouring agents as

$$\zeta_i(t) = \sum_{j=1}^{N} a_{ij} (x_i(t) - x_j(t)), \forall i = 2, 3, \cdots, N.$$

From the relationship between $\mathcal{A}$ and $\mathcal{L}$, it is easy to see that $\zeta_i(t) = \sum_{j=2}^{N} l_{ij}\xi_j(t)$. The disturbance estimation and rejection proposed in this chapter will be designed based on relative state information $\zeta_i(t)$.

6.2.1 Controller and Observer Design

The control input for disturbance rejection is designed as follows:

$$u_i(t) = cK\chi_i(t) - Fe^{Sh}\hat{\omega}_i(t), \tag{6.6}$$

where $\chi_i(t)$ and $\hat{\omega}_i(t)$ are generated by

$$\chi_i(t) = e^{Ah}\zeta_i(t) + \sum_{j=2}^{N} l_{ij} \int_{t-h}^{t} e^{A(t-\tau)} cBK\chi_j(t)\mathrm{d}\tau, \tag{6.7}$$

$$\hat{\omega}_i(t) = \eta_i(t) + L\zeta_i(t), \tag{6.8}$$

$$\dot{\eta}_i(t) = S\left(\eta_i(t) + L\zeta_i(t)\right) - LBF\sum_{j=2}^{N} l_{ij}\left(\eta_j(t) + L\zeta_j(t)\right)$$
$$- LA\zeta_i(t) - LB\sum_{j=2}^{N} l_{ij}u_j(t-h), \tag{6.9}$$

where $c \geq 2q_{\max}/r_0$ is a positive real constant with $q_{\max} = \max\{q_2, q_3, \cdots, q_N\}$, K and L are constant gain matrices to be designed later.

Remark 6.2.1 *The integral term of $\chi_i(t)$ is added in the controller design to offset the adverse effect of the time delay. If the nonlinear and disturbance terms in* (6.1) *are absent, $\chi_i(t)$ is an ideal predictor of the relative state information of the ith agent. Due to the presence of disturbance, it is a non-ideal prediction of the relative state information. Furthermore,* (6.8)*–*(6.9) *are referred to as a distributed predictor-based consensus disturbance observer, which is only dependent on the relative state information, and is independent of the information of the local state.*

Let $z_i(t) = \omega_i(t) - \hat{\omega}_i(t)$. A direct evaluation gives that

$$\dot{z}_i(t) = S\omega_i(t) - \dot{\eta}_i(t) - L\sum_{j=2}^{N} l_{ij}\dot{\xi}_j(t)$$
$$= Sz_i(t) - L\sum_{j=2}^{N} l_{ij}\psi_j - LBF\sum_{j=2}^{N} l_{ij}z_j(t), \tag{6.10}$$

which can be written in the compact form as

$$\dot{z}(t) = (I_{N-1}\otimes S)z(t) - (\mathcal{L}_1 \otimes LBF)z(t) - (\mathcal{L}_1\otimes L)\Psi, \tag{6.11}$$

where $\Psi = \left[\psi_2^T, \psi_3^T, \cdots, \psi_N^T\right]^T$.

With the control input (6.6), the closed-loop dynamics of each agent in (6.5) can be written as

$$\dot{\xi}_i(t) = A\xi_i(t) + \psi_i + BFe^{Sh}z_i(t-h) + cBKe^{Ah}\sum_{j=2}^{N} l_{ij}\xi_j(t-h)$$
$$+ cBK\sum_{j=2}^{N} l_{ij}\int_{t-h}^{t} e^{A(t-\tau)}cBK\chi_j(\tau-h)\mathrm{d}\tau, \tag{6.12}$$

where we have used $\omega_i(t) = e^{Sh}\omega_i(t-h)$ and $D = BF$.

From the error dynamics (6.5), we have

$$\xi_i(t) = e^{Ah}\xi_i(t-h) + \int_{t-h}^{t} e^{A(t-\tau)}\left(\psi_i + Bu_i(\tau-h) + D\omega_i(\tau)\right)\mathrm{d}\tau. \tag{6.13}$$

Invoking (6.13) into (6.12), we obtain

$$\begin{aligned}\dot{\xi}_i(t) =& A\xi_i(t) + cBK\sum_{j=2}^{N} l_{ij}\xi_j(t) + \psi_i + BFe^{Sh}z_i(t-h) \\ &- cBK\sum_{j=2}^{N} l_{ij}\int_{t-h}^{t} e^{A(t-\tau)}\left(\psi_j + BFe^{Sh}z_j(\tau-h)\right)\mathrm{d}\tau. \end{aligned} \tag{6.14}$$

Let $\xi = \left[\xi_2^T, \xi_3^T, \cdots, \xi_N^T\right]^T$, $z = \left[z_2^T, z_3^T, \cdots, z_N^T\right]^T$. The error dynamics of $\xi(t)$ can be written in the compact form as

$$\dot{\xi}(t) = (I\otimes A + c\mathcal{L}_1 \otimes BK)\,\xi(t) + \Psi + \left(I\otimes BFe^{Sh}\right)z(t-h) + \Delta_1 + \Delta_2, \tag{6.15}$$

where

$$\begin{aligned}\Delta_1 =& -(c\mathcal{L}_1 \otimes BK)\int_{t-h}^{t}\left(I\otimes e^{A(t-\tau)}\right)\Psi\mathrm{d}\tau, \\ \Delta_2 =& -(c\mathcal{L}_1 \otimes BK)\int_{t-h}^{t}\left(I\otimes e^{A(t-\tau)}BFe^{Sh}\right)z(\tau-h)\mathrm{d}\tau.\end{aligned}$$

For the convenience, let $\Delta_1 = \left[\delta_2^T, \delta_3^T, \cdots, \delta_N^T\right]^T$ and $\Delta_2 = \left[\bar{\delta}_2^T, \bar{\delta}_3^T, \cdots, \bar{\delta}_N^T\right]^T$.

Next, we will design the control gain K and the observer gain L. With the control law shown in (6.6), K and L are chosen as

$$K = -B^T P, \tag{6.16}$$

$$L = cQ^{-1}D^T, \tag{6.17}$$

where $P > 0, Q > 0$ are constant matrices to be designed.

To obtain the main results, the bounds on $\|\Delta_1\|^2$ and $\|\Delta_2\|^2$ are given in the following lemma.

Lemma 6.2.1 *For the terms Δ_1 and Δ_2 in the error dynamics (6.15), bounds can be established as*

$$\|\Delta_1\|^2 \le \rho_1 \int_{t-h}^{t} \xi^T(\tau)\xi(\tau)\mathrm{d}\tau, \tag{6.18}$$

$$\|\Delta_2\|^2 \le \rho_2 \int_{t-h}^{t} z^T(\tau-h)z(\tau-h)\mathrm{d}\tau, \tag{6.19}$$

where

$$\begin{aligned}\rho_1 =& (N-1)\,c^2 h\rho^2 e^{\alpha_2 h}\gamma^2 \left\|\mathcal{L}_1\right\|_F^2, \\ \rho_2 =& (N-1)h\alpha_1 c^2\rho^2 e^{(\alpha_0+\alpha_2)h}\|\mathcal{L}_1\|_F^2,\end{aligned}$$

with ρ, α_0, α_1 and α_2 *being positive numbers such that*

$$\rho^2 I \geq PBB^T BB^T P, \tag{6.20}$$
$$\alpha_0 > \lambda_{\max}(S + S^T), \tag{6.21}$$
$$\alpha_1 \geq \lambda_{\max}(F^T B^T BF), \tag{6.22}$$
$$\alpha_2 > \lambda_{\max}(A + A^T). \tag{6.23}$$

Proof 19 *From the definition of* Δ_1 *in (6.15), we have* $\|\Delta_1\|^2 = \sum_{i=2}^N \|\delta_i\|^2$. *With (6.16), we can get*

$$\delta_i = cBB^T P \sum_{j=2}^N l_{ij} \int_{t-h}^t e^{A(t-\tau)} \psi_j \mathrm{d}\tau,$$

and

$$\begin{aligned} \|\delta_i\|^2 = c^2 \int_{t-h}^t \left(\sum_{j=2}^N l_{ij} \psi_j^T \right) e^{A^T(t-\tau)} \mathrm{d}\tau PBB^T \\ \times BB^T P \int_{t-h}^t e^{A(t-\tau)} \left(\sum_{j=2}^N l_{ij} \psi_j \right) \mathrm{d}\tau. \end{aligned}$$

Based on Lemma 2.1.2 and the condition (6.20), one obtains

$$\|\delta_i\|^2 \leq c^2 h \rho^2 \int_{t-h}^t \sum_{j=2}^N l_{ij} \psi_j^T e^{(A^T + A)(t-\tau)} \sum_{j=2}^N l_{ij} \psi_j \mathrm{d}\tau.$$

In light of Lemma 2.1.1 and the condition (6.23), one gets that

$$\begin{aligned} \|\delta_i\|^2 \leq & \ (N-1)\, c^2 h \rho^2 e^{\alpha_2 h} \sum_{j=2}^N l_{ij}^2 \int_{t-h}^t \|\phi(x_j) - \phi(x_1)\|^2 \mathrm{d}\tau \\ \leq & \ (N-1)\, c^2 h \rho^2 e^{\alpha_2 h} \gamma^2 \sum_{j=2}^N l_{ij}^2 \int_{t-h}^t \xi_j^T(\tau) \xi_j(\tau) \mathrm{d}\tau \\ \leq & (N-1)\, c^2 h \rho^2 e^{\alpha_2 h} \gamma^2 \|l_i\|^2 \int_{t-h}^t \xi^T(\tau) \xi(\tau) \mathrm{d}\tau. \end{aligned}$$

Consequently,

$$\begin{aligned} \|\Delta_1\|^2 \leq & (N-1)\, c^2 h \rho^2 e^{\alpha_2 h} \gamma^2 \sum_{i=2}^N \|l_i\|^2 \int_{t-h}^t \xi^T(\tau) \xi(\tau) \mathrm{d}\tau \\ \leq & (N-1)\, c^2 h \rho^2 e^{\alpha_2 h} \gamma^2 \|\mathcal{L}_1\|_F^2 \int_{t-h}^t \xi^T(\tau) \xi(\tau) \mathrm{d}\tau. \end{aligned}$$

In a similar way, we have

$$\bar{\delta}_i = cBB^T P \sum_{j=2}^{N} l_{ij} \int_{t-h}^{t} e^{A(t-\tau)} BFe^{Sh} z_j(\tau - h) \mathrm{d}\tau.$$

It follows that

$$\begin{aligned}\|\bar{\delta}_i\|^2 &\leq c^2\rho^2 h \int_{t-h}^{t} \left(\sum_{j=2}^{N} l_{ij} z_j^T(\tau - h) \right) e^{S^T h} F^T B^T e^{A^T(t-\tau)} \\ &\quad \times e^{A(t-\tau)} BFe^{Sh} \left(\sum_{j=2}^{N} l_{ij} z_j(\tau - h) \right) \mathrm{d}\tau \\ &\leq h\alpha_1 c^2 \rho^2 e^{(\alpha_0+\alpha_2)h} \int_{t-h}^{t} \sum_{j=2}^{N} l_{ij} z_j^T(\tau - h) \sum_{j=2}^{N} l_{ij} z_j(\tau - h) \mathrm{d}\tau \\ &\leq (N-1) h\alpha_1 c^2 \rho^2 e^{(\alpha_0+\alpha_2)h} \sum_{j=2}^{N} l_{ij}^2 \int_{t-h}^{t} z_j^T(\tau - h) z_j(\tau - h) \mathrm{d}\tau.\end{aligned}$$

Consequently,

$$\begin{aligned}\|\Delta_2\|^2 &\leq (N-1) h\alpha_1 c^2 \rho^2 e^{(\alpha_0+\alpha_2)h} \sum_{i=2}^{N} \|l_i\|^2 \int_{t-h}^{t} z^T(\tau - h) z(\tau - h) \mathrm{d}\tau \\ &\leq (N-1) h\alpha_1 c^2 \rho^2 e^{(\alpha_0+\alpha_2)h} \|\mathcal{L}_1\|_F^2 \int_{t-h}^{t} z^T(\tau - h) z(\tau - h) \mathrm{d}\tau.\end{aligned}$$

This completes the proof.

6.2.2 Consensus Analysis

The following theorem presents sufficient conditions to ensure that the consensus disturbance rejection problem is solved by using the control algorithm (6.6) with the control gain K and the observer gain L in (6.16)-(6.17).

Theorem 11 *For multi-agent systems (6.1)-(6.2) with Assumptions 6.1.1 and 6.1.2, the consensus disturbance rejection problem can be solved by the control algorithm (6.6) with (6.16)-(6.17) if there exists positive definite matrices* P, Q *and constants* $\omega_1 \geq 0$, $\rho, \kappa_i > 0, i = 1, 2, \cdots, 5$, *such that*

$$\rho W - BB^T \geq 0, \tag{6.24}$$

$$\begin{bmatrix} AW + WA^T - 2BB^T + H & W \\ W & -\epsilon_1^{-1} \end{bmatrix} < 0, \tag{6.25}$$

$$QS + S^T Q - 2D^T D + \epsilon_2 I < 0, \tag{6.26}$$

are satisfied with $W = P^{-1}$ *and*

$$
\begin{aligned}
H &= (\kappa_1 + \kappa_2 + \kappa_3 + \kappa_4) I, \\
\epsilon_1 &= \left(\kappa_1^{-1} + c\kappa_5^{-1}\sigma_{\max}^2(\mathcal{L}_1)\right)\gamma^2 + \rho_1 \pi_{\min}^{-1}\pi_{\max}\kappa_3^{-1}e^h, \\
\epsilon_2 &= \pi_{\max}\pi_{\min}^{-1}\left(\alpha_1\kappa_2^{-1}e^{(\alpha_0+1)h} + c\kappa_5\lambda_{\max}(D^TD) + e^h\kappa_4^{-1}\rho_2\right),
\end{aligned}
$$

where $\pi_{\min} = \min\{\pi_2, \pi_3, \cdots, \pi_N\}$, $\pi_{\max} = \max\{\pi_2, \pi_3, \cdots, \pi_N\}$.

Proof 20 *To start the consensus analysis, we try a Lyapunov function candidate*

$$
V_0 = \xi^T (G \otimes P)\xi + z^T (G \otimes Q) z + \sigma_0 e^h \int_{t-h}^{t} e^{\tau - t} z^T(\tau) z(\tau) \mathrm{d}\tau, \tag{6.27}
$$

where σ_0 *is a positive value to be chosen later.*

The derivative of V_0 *along the trajectory of (6.11) and (6.15) can be obtained as*

$$
\begin{aligned}
\dot{V}_0 = {} & \xi^T \left(G \otimes (PA + A^TP) - c\left(G\mathcal{L}_1 + \mathcal{L}_1^TG\right) \otimes PBB^TP\right)\xi \\
& + 2\sum_{i=2}^{N} \frac{1}{\pi_i}\xi_i^T P\left(BFe^{Sh}z_i(t-h) + \psi_i + \delta_i + \bar{\delta}_i\right) \\
& + z^T(t)\left(G \otimes (QS + S^TQ) - c(G\mathcal{L}_1 + \mathcal{L}_1^TG) \otimes D^TD\right)z(t) \\
& - 2cz^T(t)(G\mathcal{L}_1 \otimes D^T)\Psi - \sigma_0 e^h \int_{t-h}^{t} e^{\tau-t}z^T(\tau)z(\tau)\mathrm{d}\tau \\
& + \sigma_0 e^h z^T(t)z(t) - \sigma_0 z^T(t-h)z(t-h) \\
\le {} & \xi^T\left(G \otimes (PA + A^TP + (\kappa_1+\kappa_2+\kappa_3+\kappa_4)PP) - cr_0 I \otimes PBB^TP\right)\xi \\
& + \frac{\gamma^2}{\kappa_1}\sum_{i=2}^{N}\frac{1}{\pi_i}\xi_i^T\xi_i + \frac{\alpha_1}{\kappa_2}e^{\alpha_0 h}\sum_{i=2}^{N}\frac{1}{\pi_i}z_i^T(t-h)z_i(t-h) + \frac{\|\Delta_1\|^2}{\kappa_3\pi_{\min}} + \frac{\|\Delta_2\|^2}{\kappa_4\pi_{\min}} \\
& + z^T(t)\left(G \otimes \left(QS + S^TQ + \frac{\pi_{\max}}{\pi_{\min}}c\kappa_5\lambda_{\max}(D^TD)I\right) - cr_0 I \otimes D^TD\right)z(t) \\
& - \sigma_0 z^T(t-h)z(t-h) + \sigma_0 e^h z^T(t)z(t) + \frac{c\gamma^2}{\kappa_5}\sigma_{\max}^2(\mathcal{L}_1)\sum_{i=2}^{N}\frac{1}{\pi_i}\xi_i^T\xi_i \\
\le {} & \xi^T\left(G \otimes (PA + A^TP - 2PBB^TP + (\kappa_1+\kappa_2+\kappa_3+\kappa_4)PP + \sigma_1 I)\right)\xi \\
& + z^T(t)\left(G \otimes (QS + S^TQ - 2D^TD + \sigma_{11}I)\right)z(t) + \frac{\|\Delta_1\|^2}{\kappa_3\pi_{\min}} + \frac{\|\Delta_2\|^2}{\kappa_4\pi_{\min}} \\
& + \left(\alpha_1\kappa_2^{-1}\pi_{\min}^{-1}e^{\alpha_0 h} - \sigma_0\right)z^T(t-h)z(t-h),
\end{aligned} \tag{6.28}
$$

where $\sigma_1 = \left(\kappa_1^{-1} + c\kappa_5^{-1}\sigma_{\max}^2(\mathcal{L}_1)\right)\gamma^2$, $\sigma_{11} = \sigma_0\pi_{\max}e^h + c\kappa_5\pi_{\max}\pi_{\min}^{-1}\lambda_{\max}(D^TD)$. *Lemmas 6.1.1, 2.1.1 and 2.1.2 are used in above derivation.*

By choosing $\sigma_0 = \alpha_1 \kappa_2^{-1} \pi_{\min}^{-1} e^{\alpha_0 h}$, *the derivative of* V_0 *could be written as*

$$\begin{aligned}
\dot{V}_0 \leq\, & \xi^T \left(G \otimes \left(PA + A^T P - 2PBB^T P + (\kappa_1 + \kappa_2 + \kappa_3 + \kappa_4) PP + \sigma_1 I\right)\right) \xi \\
& + z^T \left(G \otimes \left(QS + S^T Q - 2D^T D + \sigma_2 I\right)\right) z + \frac{\rho_1}{\kappa_3 \pi_{\min}} \int_{t-h}^{t} \xi^T(\tau)\xi(\tau)\mathrm{d}\tau \\
& + \frac{\rho_2}{\kappa_4 \pi_{\min}} \int_{t-h}^{t} z^T(\tau - h) z(\tau - h)\mathrm{d}\tau,
\end{aligned} \tag{6.29}$$

where $\sigma_2 = \pi_{\max} \pi_{\min}^{-1} \left(\alpha_1 \kappa_2^{-1} e^{(\alpha_0+1)h} + c\kappa_5 \lambda_{\max}(D^T D)\right)$.

To deal with the first integral term shown in (6.29), we consider the following Krasovskii functional

$$V_1 = e^h \int_{t-h}^{t} e^{\tau - t} \xi^T(\tau)\xi(\tau)\mathrm{d}\tau,$$

with the direct calculations as

$$\begin{aligned}
\dot{V}_1 =& - e^h \int_{t-h}^{t} e^{\tau-t}\xi^T(\tau)\xi(\tau)\mathrm{d}\tau + e^h \xi^T(t)\xi(t) - \xi^T(t-h)\xi(t-h) \\
\leq& - \int_{t-h}^{t} \xi^T(\tau)\xi(\tau)\mathrm{d}\tau + e^h \xi^T(t)\xi(t).
\end{aligned} \tag{6.30}$$

Similarly, the second integral term in (6.29) is coped with as

$$V_2 = e^h \int_{t-h}^{t} z^T(\tau) z(\tau)\mathrm{d}\tau + e^h \int_{t-h}^{t} e^{\tau-t} z^T(\tau - h) z(\tau - h)\mathrm{d}\tau,$$

with the derivative as

$$\begin{aligned}
\dot{V}_2 =& - e^h \int_{t-h}^{t} e^{\tau-t} z^T(\tau-h) z(\tau-h)\mathrm{d}\tau + e^h z^T(t) z(t) - z^T(t-2h) z(t-2h) \\
\leq& - \int_{t-h}^{t} z^T(\tau-h) z(\tau-h)\mathrm{d}\tau + e^h z^T(t) z(t).
\end{aligned} \tag{6.31}$$

Let

$$V = V_0 + \rho_1 \pi_{\min}^{-1} \kappa_3^{-1} V_1 + \rho_2 \pi_{\min}^{-1} \kappa_4^{-1} V_2. \tag{6.32}$$

A direct evaluation gives that

$$\dot{V} \leq \xi^T(t) \left(G \otimes P_1\right) \xi(t) + z^T(t) \left(G \otimes Q_1\right) z(t), \tag{6.33}$$

where

$$P_1 = PA + A^T P - 2PBB^T P + (\kappa_1 + \kappa_2 + \kappa_3 + \kappa_4) PP + \epsilon_1 I, \tag{6.34}$$

$$Q_1 = QS + S^T Q - 2D^T D + \epsilon_2 I. \tag{6.35}$$

The condition in (6.24) *is equivalent to the condition specified in* (6.20). *With* (6.34) *and* (6.35), *it can be shown by Schur Complement that conditions* (6.25) *and* (6.26) *are respectively equivalent to* $P_1 < 0$ *and* $Q_1 < 0$, *which further implies from* (6.33) *that* $\dot{V}(t) < 0$. *Thus, the error dynamics systems* (6.5) *are globally asymptotically stable at the origin, which implies that the consensus disturbance rejection of the multi-agent systems (6.1)–(6.2) is achieved. This completes the proof.*

Remark 6.2.2 *It is observed that (6.25) and (6.26) are more likely to be satisfied if the values of* $\omega_1, \rho, \kappa_i, i = 1, 2, \cdots, 5$ *are small. Therefore, the step by step algorithm designed in [162] for finding a feasible solution of the conditions may also be applied here. Additionally, since the values of* h *and* γ *are fixed and they are not the decision variables of the LMIs, a feasible solution may not exist if the values of* h *and* γ *are too large.*

6.3 A Numerical Example

In this section, we will demonstrate the consensus disturbance rejection method under the leader-follower setup of five subsystems subject to the connection topology specified by the following adjacency matrix

$$\mathcal{A} = \begin{bmatrix} 0 & 0 & 0 & 0 & 0 \\ 1 & 0 & 0 & 1 & 0 \\ 0 & 1 & 0 & 0 & 0 \\ 0 & 0 & 1 & 0 & 1 \\ 1 & 1 & 0 & 0 & 0 \end{bmatrix}.$$

Note that the first row all are zeros, as the agent indexed by 1 is taken as the leader. The communication graph is shown in Figure 6.1, from which it shows that only the followers indexed by 2 and 5 can get access to the leader and the communication topology contains a directed spanning tree. The dynamics of the ith agent are described by (6.1), with

$$A = \begin{bmatrix} 0 & -1 \\ 1 & 0 \end{bmatrix}, \quad B = \begin{bmatrix} 1 & 0.5 \\ 0.5 & 1 \end{bmatrix}, \phi(x_i) = g \begin{bmatrix} \sin(x_{i1}(t)) \\ \sin(x_{i2}(t)) \end{bmatrix},$$

which may present a practical dynamical model of unmanned aerial vehicle (UAV) [181]. In this scenario, it is supposed that external disturbance and time delay exist in the control channel. The external disturbance $w_i(t)$ is generated by (6.3) with

$$S = \begin{bmatrix} 0 & -0.1 \\ 0.1 & 0 \end{bmatrix}, \quad F = \begin{bmatrix} 1 & -0.5 \\ -0.5 & 1 \end{bmatrix},$$

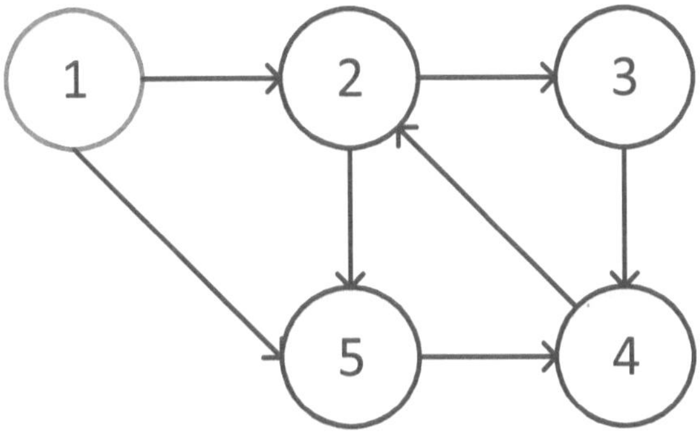

FIGURE 6.1: Communication topology.

which represents an external periodic disturbance with known frequency but without any information of its magnitude and phase. The input delay of each agent is $0.1s$, and the Lipschitz constant is $\gamma = g = 0.05$. It can be checked that both Assumptions 6.1.1 and 6.1.2 are satisfied.

The Laplacian matrix $\mathcal{L}_1$ associated with $\mathcal{A}$ is that

$$\mathcal{L}_1 = \begin{bmatrix} 2 & 0 & -1 & 0 \\ -1 & 1 & 0 & 0 \\ 0 & -1 & 2 & -1 \\ -1 & 0 & 0 & 2 \end{bmatrix}.$$

Following Lemma 6.1.1, we obtain that $G = \text{diag}\{0.3846 \;\; 0.3571 \;\; 0.5556 \;\; 0.7143\}$ and $r_0 = 0.2573$. With $p_{\max} = 0.7143$ and $2p_{\max}/r_0 = 5.5523$, we set $c = 6$ in the control input (6.6).

The initial states of agents are chosen randomly within $[\ 0,\ 5\ ]$, and $u(\theta) = [\ 0,\ 0,\ 0,\ 0\ ]^T,\ \forall\theta \in [-h, 0]$. With $\rho = 0.2$, feasible solutions of the feedback gain K and the observer gain L are found to be

$$K = \begin{bmatrix} -0.1924 & -0.1233 \\ -0.0962 & -0.2466 \end{bmatrix}, \quad L = \begin{bmatrix} 11.2500 & 0 \\ 0 & 11.2500 \end{bmatrix}.$$

Simulation study has been carried out with different disturbances for agents. Figures 6.2 and 6.3 show the simulation results for the trajectories of the states. The tracking errors between the four followers and the leader are shown in Figures 6.4 and 6.5. The disturbance observation errors are shown in Figure 6.6. From the results shown in these figures, it can be seen clearly that all the five agents reach consensus although they are under different disturbances. Therefore, the conditions specified in Theorem 11 are sufficient to guarantee the consensus disturbance rejection.

Moreover, as only the Lipschitz constant γ is used for the disturbance observer design and the exact information of the nonlinear functions is not required, this leads to conservatism in the presented conditions. With the

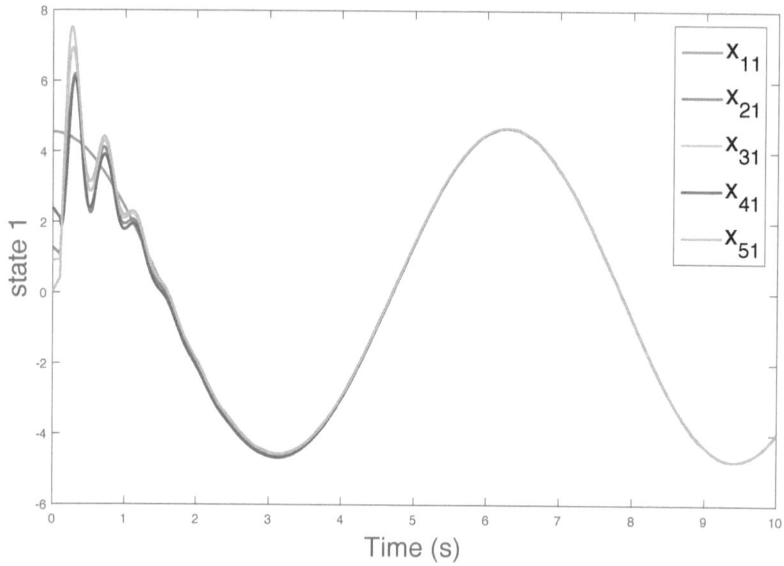

FIGURE 6.2: The trajectories of state 1 with $h = 0.1$ and $g = 0.05$.

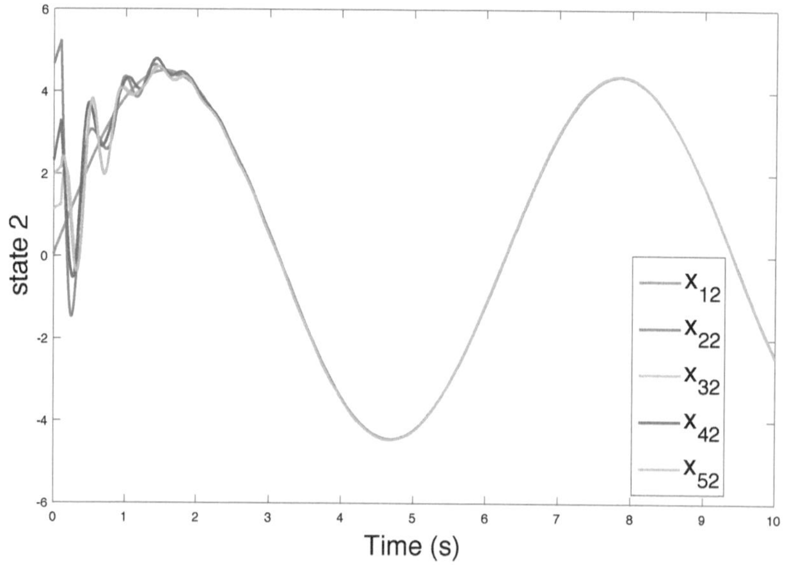

FIGURE 6.3: The trajectories of state 2 with $h = 0.1$ and $g = 0.05$.

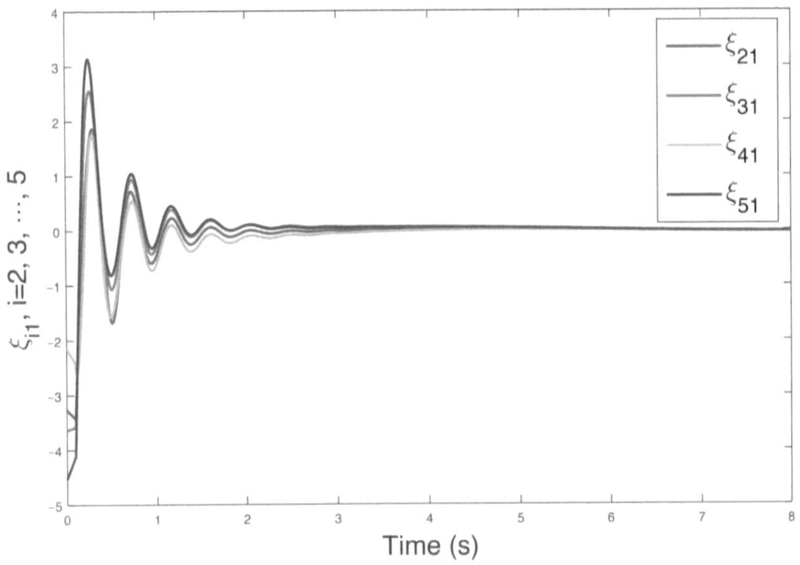

FIGURE 6.4: The evolutions of tracking errors $x_{i1} - x_{11}$.

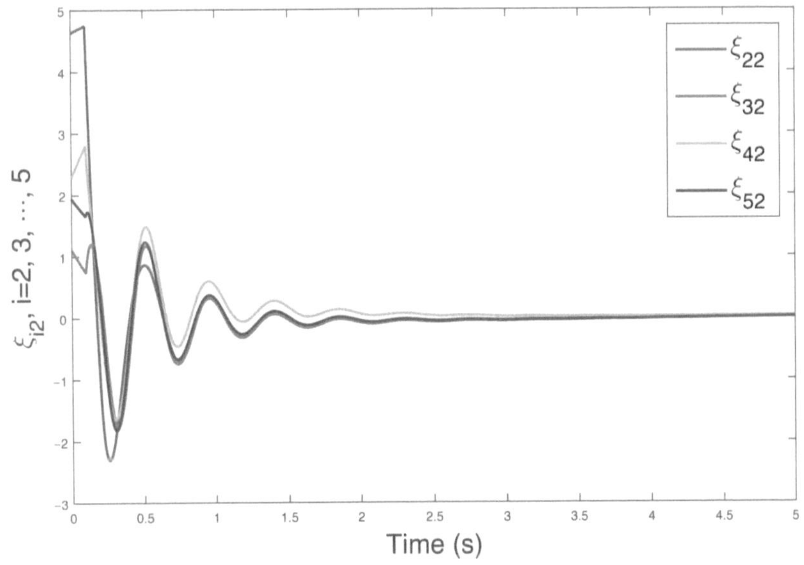

FIGURE 6.5: The evolutions of tracking errors $x_{i2} - x_{12}$.

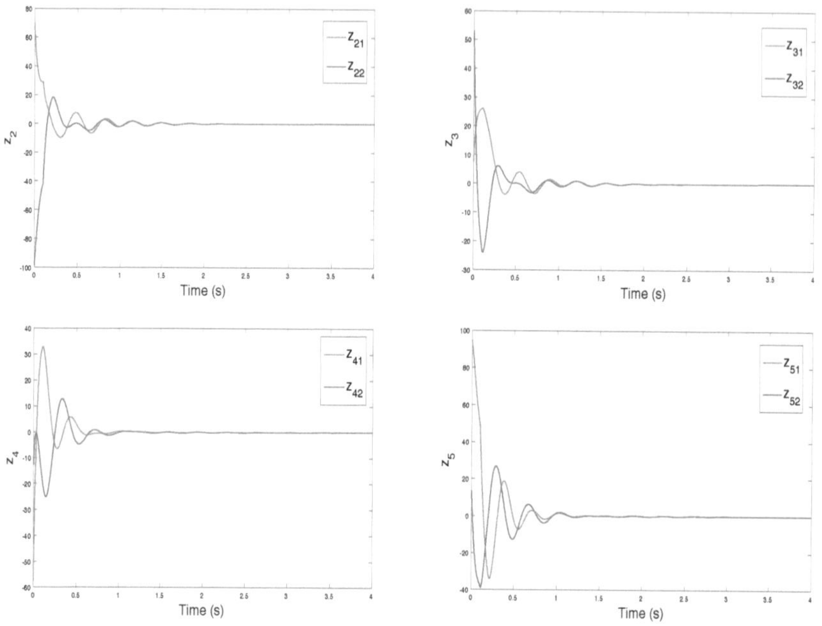

FIGURE 6.6: The estimation errors of the disturbance observers.

same control gain, the consensus disturbance rejection is still achieved for the multi-agent systems with a much larger Lipschitz constant.

6.4 Conclusions

In this chapter, we have addressed the consensus disturbance rejection problem for Lipschitz nonlinear multi-agent systems with input delay under the directed communication graph. The input delay may represent some delays in the network communication or in the actuators. Based on the conventional predictor-based feedback approach, a non-ideal predictor-based control scheme is constructed for each subsystem by using the estimate of the disturbance and the information of the relative state. By exploring certain features of the Laplacian matrix, global consensus analysis is put in the framework of Lyapunov analysis. The proposed analysis ensures that the integral terms of the system state are carefully considered by using Krasovskii functionals. Sufficient conditions are derived for the Lipschitz nonlinear systems with input delay to guarantee consensus with disturbance rejection in the time domain.

Finally, an example is employed to demonstrate the validity of the theoretical results.

6.5 Notes

This chapter extends an idea in [26, 161] and develops consensus protocols for Lipschitz nonlinear multi-agent systems with input delay and external disturbances. The main materials of this chapter are based on [164].

Chapter 7

Consensus Disturbance Rejection for Lipschitz Nonlinear MASs with Input Delay: A Predictive Observation Approach

In this chapter, we study output feedback leader-follower consensus problem for multi-agent systems subject to external disturbances and time delays in both input and output. First, we consider the linear case and a novel predictor-based extended state observer (ESO) is designed for each follower with relative output information of the neighbouring agents. Then, leader-follower consensus protocols are proposed which can compensate the delays and disturbances efficiently. In particular, the proposed observer and controller do not contain any integral term of the past control input and hence are easy to implement. Consensus analysis is put in the framework of Lyapunov-Krasovskii functionals and sufficient conditions are derived to guarantee that the consensus errors converge to zero asymptotically. Then, the results are extended to nonlinear multi-agent systems with nonlinear disturbances. Finally, the validity of the proposed design is demonstrated through a numerical example of network-connected UAVs.

7.1 Problem Formulation

Consider a group of $N+1$ agents consisting of N followers and one leader indexed by 0 (here and hereafter, the argument t is omitted excepting delayed arguments)

$$\begin{cases} \dot{x}_i = Ax_i + Bu_i(t-\tau_u) + BF\omega_i \\ y_i = Cx_i(t-\tau_y) \\ \dot{\omega}_i = S\omega_i \end{cases}, \tag{7.1}$$

where for agent i, $i = 0, 1, \ldots, N$, $x_i \in \mathbb{R}^n$ is the state, $u_i \in \mathbb{R}^q$ is the control input, $y_i \in \mathbb{R}^p$ is the output. $A \in \mathbb{R}^{n\times n}$, $B \in \mathbb{R}^{n\times q}$, $C \in \mathbb{R}^{p\times n}$ and $F \in \mathbb{R}^{q\times s}$ are constant matrices with (A, B) being controllable and (A, C) being observable, $\tau_u, \tau_y > 0$ are the input and the output delay, respectively,

$S \in \mathbb{R}^{s \times s}$ is a known constant matrix, $\omega_i \in \mathbb{R}^s$ is external input disturbance with unknown bound. For the leader-follower structure, it is reasonable to assume that the leader has no neighbours, and the leader's control input is zero [29, 79], i.e. $u_0 \equiv 0$, $\omega_0 \equiv 0$.

The consensus disturbance rejection problem considered in this chapter is to design an observer-based control algorithm for each follower by using its relative output information such that the system (7.1) can reach state consensus with disturbance rejection. That is, under the proposed algorithm, the following hold for all initial conditions

$$\lim_{t \to \infty} (x_i - x_0) = 0, \quad \forall i = 1, 2, \cdots, N. \tag{7.2}$$

Assumption 7.1.1 *The eigenvalues of S are distinct and on the imaginary axis. Furthermore, the pair (S, BF) is observable.*

Assumption 7.1.2 *The communication topology $\mathcal{G}$ contains a directed spanning tree with the leader as the root.*

Remark 7.1.1 *From a practical point of view, any periodic disturbances can be approximated by sinusoidal functions with different frequencies, and those sinusoidal functions can be formulated as the state variables of the exosystem under Assumption 7.1.1 [28]. For the convenience of presentation, in this chapter, it is assumed that the disturbance frequency is same for all the agents which determined by S matrix. With some slight adjustments, the results can be extended to $\dot{\omega}_i = S_i \omega_i$ case as shown in [143].*

7.2 Predictive Observer-Based Consensus for Linear Case

7.2.1 Predictor-Based ESO and Controller Design

Let $\eta_i = x_i - x_0$. Then, we have

$$\begin{cases} \dot{\eta}_i = A\eta_i + Bu_i(t - \tau_u) + BF\omega_i \\ \tilde{y}_i = C\eta_i(t - \tau_y) \end{cases}, \tag{7.3}$$

where $\tilde{y}_i = y_i - y_0$.

Define a new state $z_i = [\eta_i^T, \omega_i^T]^T$, which includes the exosystem model. The state-space equation (7.1) can be rewritten in the augmented form

$$\begin{cases} \dot{z}_i = A_z z_i + B_z u_i(t - \tau_u) \\ \tilde{y}_i = C_z z_i(t - \tau_y) \end{cases}, \tag{7.4}$$

where

$$A_z = \begin{bmatrix} A & BF \\ 0 & S \end{bmatrix}, B_z = \begin{bmatrix} B \\ 0 \end{bmatrix}, C_z = \begin{bmatrix} C & 0 \end{bmatrix}.$$

Define $\bar{z}_i = [\bar{\eta}_i^T, \bar{\omega}_i^T]^T$ as the estimation of the state z_i at time $t + \tau_u$. Then, a predictor-based extended-state-observer is constructed as

$$\dot{\bar{z}}_i = A_z \bar{z}_i + B_z u_i + L \sum_{j=1}^{N} l_{ij} \left(\tilde{y}_j - C_z \bar{z}_j (t - \tau) \right), \tag{7.5}$$

where $\tau = \tau_u + \tau_y$, L is the observer gain matrix to be designed later. The estimation error is defined by $\tilde{e}_i = [\tilde{e}_{\eta_i}^T, \tilde{e}_{\omega i}^T]^T = z_i - \bar{z}_i(t - \tau_u)$. Then, we have

$$\begin{aligned} \dot{\tilde{e}}_i &= A_z \tilde{e}_i - L \sum_{j=1}^{N} l_{ij} \left(C_z z_j (t - \tau) - C_z \bar{z}_j (t - \tau - \tau_u) \right) \\ &= A_z \tilde{e}_i - L C_z \sum_{j=1}^{N} l_{ij} \tilde{e}_j (t - \tau) \\ &= A_z \tilde{e}_i - L C_z v_i (t - \tau), \end{aligned} \tag{7.6}$$

where $v_i = -\sum_{k=1}^{N} l_{ik} \tilde{e}_k$. From the error dynamics (7.6), we have

$$\tilde{e}_i = e^{A_z \tau} \tilde{e}_i (t - \tau) + \int_{t-\tau}^{t} e^{A_z (t-s)} L C_z v_i (s - \tau) \mathrm{d}s, \tag{7.7}$$

and

$$\begin{aligned} \dot{\tilde{e}}_i &= A_z \tilde{e}_i - L C_z e^{-A_z \tau} \sum_{j=1}^{N} l_{ij} e^{A_z \tau} \tilde{e}_j (t - \tau) \\ &= A_z \tilde{e}_i - L \bar{C}_z \sum_{j=1}^{N} l_{ij} \tilde{e}_j + L \bar{C}_z \sum_{j=1}^{N} l_{ij} \int_{t-\tau}^{t} e^{A_z (t-s)} L C_z v_j (s - \tau) \mathrm{d}s, \end{aligned}$$

where $\bar{C}_z = C_z e^{-A_z \tau}$. Define $\tilde{e} = \left[\tilde{e}_1^T, \tilde{e}_2^T, \cdots, \tilde{e}_N^T \right]^T$ and $v = \left[v_1^T, v_2^T, \cdots, v_N^T \right]^T$. The closed-loop system is then described by

$$\dot{\tilde{e}} = \left[(I \otimes A_z) - \left(\mathcal{L}_1 \otimes L \bar{C}_z \right) \right] \tilde{e} + \lambda, \tag{7.8}$$

where I represents the identity matrix with appropriate dimension, and

$$\lambda = \left(\mathcal{L}_1 \otimes L \bar{C}_z \right) \int_{t-\tau}^{t} \left(I \otimes e^{A_z (t-s)} L C_z \right) v(s - \tau) \mathrm{d}s.$$

The controller for the ith agent is designed as

$$u_i = -K \bar{\eta}_i - F \bar{\omega}_i = -[K, F] \bar{z}_i, \tag{7.9}$$

where K is the control gain matrix to be designed later. Under control algorithm (7.9), multi-agent systems (7.3) can be written as

$$\begin{aligned}\dot{\eta}_i &= A\eta_i - BK\bar{\eta}_i(t-\tau_u) - BF\bar{\omega}_i(t-\tau_u) + BF\omega_i \\ &= (A-BK)\,\eta_i + BB_1\tilde{e}_i,\end{aligned}$$

where $B_1 = [K, F]$. Let $\eta = \left[\eta_1^T, \eta_2^T, \cdots, \eta_N^T\right]^T$. The closed-loop system is then described by

$$\dot{\eta} = (I \otimes (A - BK))\,\eta + (I \otimes BB_1)\,\tilde{e}. \tag{7.10}$$

Remark 7.2.1 *Unlike [57], which requires the state of the leader to be accessed by all the followers, in this chapter we assume that only a few followers can access the output information of the leader. Under this assumption,* $\tilde{y}_j = y_j - y_0$ *in (7.5) is not available to all the agents. However, based on the relative output information of the neighbouring agent, we have*

$$\begin{aligned}\sum_{j=0}^{N} a_{ij}\,(y_i - y_j) &= \sum_{j=1}^{N} a_{ij}\,(y_i - y_j) + a_{i0}(y_i - y_0) \\ &= \sum_{j=1}^{N} a_{ij}\,(\tilde{y}_i - \tilde{y}_j) + a_{i0}\tilde{y}_i = \sum_{j=1}^{N} l_{ij}\tilde{y}_j.\end{aligned}$$

From this equation, it can be verified that observer (7.5) is implementable even if some agents cannot access the leader's information.

7.2.2 Stability Analysis

With the observer and the control law shown in (7.5) and (7.9), the control and observer gains are chosen as

$$K = B^T P_1, \; L = cP_2^{-1}\bar{C}_z^T, \tag{7.11}$$

where P_1 and P_2 are positive definite matrices, c is a constant such that $c \geq 2g_{\max}/\rho_0$, $g_{\max} = \max\{g_1, g_2, \cdots, g_N\}$.

For the stability analysis, first we need to establish a bound of the extra integral term λ.

Lemma 7.2.1 *For the integral term* λ*, a bound can be established as*

$$\|\lambda\|^2 \leq \sigma_0 \int_{t-\tau}^{t} \tilde{e}^T(s-\tau)\,\tilde{e}\,(s-\tau)\,\mathrm{d}s, \tag{7.12}$$

where $\sigma_0 = \tau\,\|\mathcal{L}_1\|_F^4\,\rho^4 e^{2\alpha\tau}$*,* α *is a positive number such that* $\alpha \geq \lambda_{\max}\left(A_z^T + A_z\right)$*,* ρ *is a positive real number such that*

$$\rho^2 I \geq c^2 \bar{C}_z^T \bar{C}_z P_2^{-1} P_2^{-1} \bar{C}_z^T \bar{C}_z, \tag{7.13}$$

and $\|\cdot\|_F$ *denotes the Frobenius norm of a matrix.*

Proof 21 *Let* $\lambda = \left[\lambda_1^T, \lambda_2^T, \cdots, \lambda_N^T\right]^T$. *From (7.8), we have*

$$\lambda_i = L\bar{C}_z \sum_{j=1}^{N} l_{ij} \int_{t-\tau}^{t} e^{A_z(t-s)} LC_z v_j(s-\tau)\mathrm{d}s$$
$$= -L\bar{C}_z \sum_{j=1}^{N} l_{ij} \int_{t-\tau}^{t} e^{A_z(t-s)} LC_z \sum_{k=1}^{N} l_{jk}\tilde{e}_k(s-\tau)\mathrm{d}s.$$

We define

$$\theta_k = -L\bar{C}_z \int_{t-\tau}^{t} e^{A_z(t-s)} LC_z \tilde{e}_k(s-\tau)\mathrm{d}s.$$

Then, we obtain that

$$\lambda_i = \sum_{j=1}^{N} l_{ij} \sum_{k=1}^{N} l_{jk}\theta_k.$$

Let $\theta = \left[\theta_1^T, \theta_2^T, \cdots, \theta_N^T\right]^T$. *It then follows that*

$$\|\lambda_i\| \leq \sum_{j=1}^{N} |l_{ij}| \sum_{k=1}^{N} |l_{jk}| \|\theta_k\| \leq \|l_i\| \|\mathcal{L}_1\|_F \|\theta\|,$$

where l_i *denotes the ith row of* $\mathcal{L}_1$. *Therefore, we have*

$$\|\lambda\|^2 = \sum_{i=1}^{N} \|\lambda_i\|^2 \leq \sum_{i=1}^{N} \|l_i\|^2 \|\mathcal{L}_1\|_F^2 \|\theta\|^2 = \|\mathcal{L}_1\|_F^4 \|\theta\|^2.$$

Next, we need to deal with $\|\theta\|^2$. *By Lemma 2.1.2, we have*

$$\|\theta_i\|^2 = c^2 \int_{t-\tau}^{t} \tilde{e}_i^T(s-\tau)C_z^T L^T e^{A_z^T(t-s)}\mathrm{d}s\bar{C}_z^T\bar{C}_z P_2^{-1}$$
$$\times P_2^{-1}\bar{C}_z^T\bar{C}_z \int_{t-\tau}^{t} e^{A_z(t-s)} LC_z\tilde{e}_i(s-\tau)\mathrm{d}s$$
$$\leq \tau\rho^2 \int_{t-\tau}^{t} \tilde{e}_i^T(s-\tau)C_z^T L^T e^{A_z^T(t-s)} \times e^{A_z(t-s)} LC_z\tilde{e}_i(s-\tau)\mathrm{d}s.$$

In view of Lemma 2.1.1 with $P = I$, $\alpha \geq \lambda_{\max}\left(A_z^T + A_z\right)$, *we have* $e^{A_z^T t}e^{A_z t} \leq e^{\alpha t}I$, *and*

$$\|\theta_i\|^2 \leq \tau\rho^2 c^2 \int_{t-\tau}^{t} e^{\alpha(t-s)}\tilde{e}_i^T(s-\tau)e^{A_z^T\tau}\bar{C}_z^T\bar{C}_z P_2^{-1}P_2^{-1}\bar{C}_z^T\bar{C}_z e^{A_z\tau}\tilde{e}_i(s-\tau)\mathrm{d}s$$
$$\leq \tau\rho^4 e^{2\alpha\tau} \int_{t-\tau}^{t} \tilde{e}_i^T(s-\tau)\tilde{e}_i(s-\tau)\mathrm{d}s.$$

Then, $\|\theta\|^2$ can be bounded as

$$\|\theta\|^2 = \sum_{i=1}^{N} \|\theta_i\|^2 \leq \tau\rho^4 e^{2\alpha\tau} \int_{t-\tau}^{t} \tilde{e}^T(s-\tau)\tilde{e}(s-\tau)\mathrm{d}s.$$

Finally, we have

$$\|\lambda\|^2 \leq \tau \|\mathcal{L}_1\|_F^4 \rho^4 e^{2\alpha\tau} \int_{t-\tau}^{t} \tilde{e}^T(s-\tau)\tilde{e}(s-\tau)\mathrm{d}s.$$

This completes the proof.

Based on the above results, the following theorem presents sufficient conditions such that the consensus disturbance rejection problem is solved by using the relative output information.

Theorem 12 *For multi-agent systems (7.1) with Assumption 7.1.2, the consensus disturbance rejection problem can be solved by the observer (7.5) and the controller (7.9) with (7.11) if there exist positive definite matrices P_1, P_2 and constants $\kappa, \rho > 0$, such that*

$$AW + WA^T - BB^T < 0, \tag{7.14}$$

$$\rho P_2 - c\bar{C}_z^T\bar{C}_z \geq 0, \tag{7.15}$$

$$\begin{bmatrix} P_2A_z + A_z^TP_2 - 2\bar{C}_z^T\bar{C}_z + H & P_2 \\ P_2 & -\kappa^{-1} \end{bmatrix} < 0, \tag{7.16}$$

are satisfied with $W = P_1^{-1}$ and $H = g_{\min}^{-1}B_1^TB_1 + \sigma_1 I$, where $g_{\min} = \min\{g_1, g_2, \cdots, g_N\}$, $\sigma_1 = \kappa^{-1}g_{\max}g_{\min}^{-1}e^{\tau}\sigma_0$, $\sigma_0 = \tau\|\mathcal{L}_1\|_F^4\rho^4e^{2\alpha\tau}$ is a positive number defined in Lemma 7.2.1.

Proof 22 *To start the consensus analysis, we try a Lyapunov function candidate*

$$V_0 = \eta^T(I \otimes P_1)\eta + \tilde{e}^T(G \otimes P_2)\tilde{e}. \tag{7.17}$$

In view of (7.8) and (7.10), we have

$$\begin{aligned} \dot{V}_0 = {} & \eta^T\left[I \otimes (A^TP_1 + P_1A - 2P_1BB^TP_1)\right]\eta \\ & + \tilde{e}^T\left[G \otimes (A_z^TP_2 + P_2A_z) - c\left(G\mathcal{L}_1 + \mathcal{L}_1^TG\right) \otimes \bar{C}_z^T\bar{C}_z\right]\tilde{e} \\ & + 2\eta^T(I \otimes P_1BB_1)\tilde{e} + 2\tilde{e}^T(G \otimes P_2)\lambda \\ \leq {} & \eta^T\left[I \otimes (A^TP_1 + P_1A - P_1BB^TP_1)\right]\eta + \kappa^{-1}g_{\max}\|\lambda\|^2 \\ & + \tilde{e}^T\left[G \otimes (A_z^TP_2 + P_2A_z - 2\bar{C}_z^T\bar{C}_z + \kappa P_2P_2) + B_1^TB_1\right]\tilde{e}, \end{aligned} \tag{7.18}$$

where Lemmas 1.4.4 and 1.4.9 are used for the derivation.

Using (7.12) and (7.18), we obtain that

$$
\begin{aligned}
\dot{V}_0 \leq\ & \eta^T \left[I \otimes \left(A^T P_1 + P_1 A - P_1 B B^T P_1\right)\right] \eta \\
& + \tilde{e}^T \left[G \otimes \left(A_z^T P_2 + P_2 A_z - 2\bar{C}_z^T \bar{C}_z + \kappa P_2 P_2\right) + B_1^T B_1\right] \tilde{e} \\
& + \kappa^{-1} g_{\max} \sigma_0 \int_{t-\tau}^{t} \tilde{e}^T (s-\tau)\, \tilde{e}(s-\tau)\, \mathrm{d}s. \qquad (7.19)
\end{aligned}
$$

For the delayed term shown in (7.19), we consider the following Krasovskii functional

$$
V_1 = e^{\tau} \int_{t-\tau}^{t} e^{s-t} \tilde{e}^T (s-\tau) \tilde{e}(s-\tau) \mathrm{d}s + e^{\tau} \int_{t-\tau}^{t} \tilde{e}^T (s) \tilde{e}(s) \mathrm{d}s.
$$

A direct evaluation gives that

$$
\begin{aligned}
\dot{V}_1 &= -e^{\tau} \int_{t-\tau}^{t} e^{s-t} \tilde{e}^T (s-\tau) \tilde{e}(s-\tau) \mathrm{d}s - \tilde{e}^T (t-2\tau) \tilde{e}(t-2\tau) + e^{\tau} \tilde{e}^T \tilde{e} \\
&\leq - \int_{t-\tau}^{t} \tilde{e}^T (s-\tau) \tilde{e}(s-\tau) \mathrm{d}s + e^{\tau} \tilde{e}^T \tilde{e}. \qquad (7.20)
\end{aligned}
$$

Let

$$
V = V_0 + \kappa^{-1} g_{\max} \sigma_0 V_1. \qquad (7.21)
$$

From (7.19), (7.20) and (7.21), we obtain that

$$
\dot{V} \leq \eta^T (I \otimes H_1) \eta + \tilde{e}^T (G \otimes H_2) \tilde{e}, \qquad (7.22)
$$

where

$$
\begin{aligned}
H_1 &:= A^T P_1 + P_1 A - P_1 B B^T P_1, \qquad (7.23) \\
H_2 &:= A_z^T P_2 + P_2 A_z - 2 C_z^T C_z + \kappa P_2 P_2 + g_{\min}^{-1} B_1^T B_1 + \sigma_1 I, \qquad (7.24)
\end{aligned}
$$

with $\sigma_1 = g_{\min}^{-1} e^{\tau} \kappa^{-1} g_{\max} \sigma_0$ being a positive number defined in Theorem 12.

From the analysis in this section, we know that the control law (7.9) stabilizes η and $\tilde{e}$ if $H_1 < 0$ and $H_2 < 0$ in (7.22) are satisfied. Indeed, it is easy to see the conditions $H_1 < 0$ and $H_2 < 0$ are equivalent to the conditions specified in (7.14) and (7.16). Furthermore, the condition specified in (7.15) is equivalent to the condition (7.13). It implies that η and $\tilde{e}$ converge to zero asymptotically. Hence, the consensus with disturbance rejection in (7.2) is achieved.

Remark 7.2.2 *The conditions shown in (7.14)-(7.16) can be checked by standard LMI routines for a set of fixed values. In particular, we suggest the following step by step algorithm.*

(1) Solve the LMI equation (7.14) for W, and then $B_1 = [K, F] = [B^T P_1, F]$ is fixed with $P_1 = W^{-1}$.

(2) Fix the value of ρ, κ to some constants $\tilde{\rho}, \tilde{\kappa} > 0$; make an initial guess for the values of $\tilde{\rho}, \tilde{\kappa}$.

(3) Solve the LMI equation (7.16) for P_2 with the fixed values; if a feasible value of P_2 cannot be found, return to Step (2) and reset the values of $\tilde{\rho}, \tilde{\kappa}$.

(4) Solve the LMI equation (7.15) for ρ with the feasible value of P_2 obtained in Step (3) and make sure that the value of ρ is minimized.

(5) If the condition $\tilde{\rho} \geq \rho$ is satisfied, then $(\tilde{\rho}, \tilde{\kappa}, W, P_2)$ is a feasible solution for Theorem 1; otherwise, set $\tilde{\rho} = \rho$ and return to Step (3).

Remark 7.2.3 *As mentioned in [174], unlike low-order time-delayed linear systems, where necessary and sufficient conditions for the stability of such systems have been determined by analysing the positions of the roots of the characteristic equations, for high-order time-delayed linear systems (including the multi-agent case considered in this chapter), only sufficient conditions for the stability of such systems have been determined through Lyapunov stability analysis. The delay bound and closed-loop parameters are simultaneously involved in these sufficient conditions, which are typically in the form of LMIs. Thus, to find the upper bound of τ, we may use τ^* to replace τ in (7.14)–(7.16) and solve the following optimization problem*

$$\bar{\tau} = \sup_{P_1 \geq 0, P_2 \geq 0, \kappa \geq 0, \rho \geq 0} \{\tau^*\} \text{ s.t. } (7.14)\text{–}(7.16).$$

7.3 Predictor Observer Design for Nonlinear Case

In this section, we consider the predictor-based extended-state-observer design for nonlinear case

$$\begin{cases} \dot{x}_i = Ax_i + Bu_i(t - \tau_u) + BF\omega_i + f_1(x_i) \\ y_i = Cx_i(t - \tau_y) \\ \dot{\omega}_i = S\omega_i + f_2(\omega_i) \end{cases}, \tag{7.25}$$

where $f_1(\cdot)$ and $f_2(\cdot)$ are Lipschitz nonlinear functions with Lipschitz constants γ_1, γ_2, such that

$$\begin{cases} \|f_1(x) - f_1(y)\| \leq \gamma_1 \|x - y\| \\ \|f_2(x) - f_2(y)\| \leq \gamma_2 \|x - y\| \end{cases}. \tag{7.26}$$

Let $\xi_i = x_i - x_0$. Then, we have

$$\begin{cases} \dot{\xi}_i = A\xi_i + Bu_i(t - \tau_u) + BF\omega_i + \tilde{f}_i(x_i, x_0) \\ \bar{y}_i = C\xi_i(t - \tau_y) \end{cases}, \tag{7.27}$$

where $\bar{y}_i = y_i - y_0$, $\tilde{f}_i(x_i, x_0) = f_1(x_i) - f_1(x_0)$.

Define a new state $\chi_i = [\xi_i^T, \omega_i^T]^T$, which includes the exosystem model. The state-space equation (7.27) can be rewritten in the augmented form

$$\begin{cases} \dot{\chi}_i = A_\xi \chi_i + B_\xi u_i(t-\tau_u) + \mathcal{F}_i \\ \bar{y}_i = C_\xi \xi_i(t-\tau_y) \end{cases}, \tag{7.28}$$

where

$$A_\xi = \begin{bmatrix} A & BF \\ 0 & S \end{bmatrix}, B_\xi = \begin{bmatrix} B \\ 0 \end{bmatrix}, \mathcal{F}_i = \begin{bmatrix} \tilde{f}_i(x_i, x_0) \\ f_2(\omega_i) \end{bmatrix}, C_\xi = \begin{bmatrix} C & 0 \end{bmatrix}.$$

Define $\bar{\chi}_i = [\bar{\xi}_i^T, \bar{\omega}_i^T]^T$ as the estimation of the augmented state χ_i at time $t+\tau_u$. Then, a predictor-type observer is constructed as

$$\dot{\bar{\chi}}_i = A_\xi \bar{\chi}_i + B_\xi u_i + \bar{\mathcal{F}}_i + L \sum_{j=1}^{N} l_{ij} \left(\bar{y}_j - C_\xi \bar{\chi}_j(t-\tau) \right), \tag{7.29}$$

where $\bar{\mathcal{F}}_i = \left[0, f_2^T(\bar{\omega}_i)\right]^T$, L is the observer gain matrix. The estimation error is defined by

$$\bar{e}_i = [\bar{e}_{\xi i}^T, \bar{e}_{\omega i}^T]^T = \chi_i - \bar{\chi}_i(t-\tau_u).$$

It follows that

$$\begin{aligned} \dot{\bar{e}}_i &= A_\xi e_i - L \sum_{j=1}^{N} l_{ij} \left(C_\xi \chi_j(t-\tau) - C_\xi \bar{\chi}_j(t-\tau-\tau_u) \right) + \tilde{\mathcal{F}}_i \\ &= A_\xi \bar{e}_i - L C_\xi \sum_{j=1}^{N} l_{ij} \bar{e}_j(t-\tau) + \tilde{\mathcal{F}}_i, \end{aligned} \tag{7.30}$$

where $\tilde{\mathcal{F}}_i = \mathcal{F}_i - \bar{\mathcal{F}}_i(t-\tau_u)$. Similar to (7.7), we obtain that

$$\begin{aligned} \dot{\bar{e}}_i = {} & A_\xi \bar{e}_i - L\bar{C}_\xi \sum_{j=1}^{N} l_{ij} \bar{e}_j + \tilde{\mathcal{F}}_i \\ & + L\bar{C}_\xi \sum_{j=1}^{N} l_{ij} \int_{t-\tau}^{t} e^{A_\xi (t-s)} \left[L C_\xi \bar{v}_j(s-\tau) + \tilde{\mathcal{F}}_j \right] \mathrm{d}s, \end{aligned}$$

where $\bar{v}_j - -\sum_{k=1}^{N} l_{jk} \bar{e}_k$, $\bar{C}_\xi = C_\xi c^{-A_\xi \tau}$. Dcfinc $\bar{c} = \left[\bar{e}_1^T, \bar{e}_2^T, \cdots, \bar{e}_N^T\right]^T$, $\bar{v} = \left[\bar{v}_1^T, \bar{v}_2^T, \cdots, \bar{v}_N^T\right]^T$, and $\tilde{\mathcal{F}} = \left[\tilde{\mathcal{F}}_1^T, \tilde{\mathcal{F}}_2^T, \cdots, \tilde{\mathcal{F}}_N^T\right]^T$. The closed-loop system is then described by

$$\dot{\bar{e}} = \left[(I \otimes A_\xi) - \left(\mathcal{L}_1 \otimes L\bar{C}_\xi \right) \right] \bar{e} + \bar{\lambda} + \phi, \tag{7.31}$$

where

$$\bar{\lambda} = (\mathcal{L}_1 \otimes L\bar{C}_\xi) \int_{t-\tau}^{t} \left(I \otimes e^{A_\xi(t-s)} LC_z\right) \bar{v}(s-\tau)\mathrm{d}s,$$
$$\phi = (\mathcal{L}_1 \otimes L\bar{C}_\xi) \int_{t-\tau}^{t} \left(I \otimes e^{A_\xi(t-s)}\right) \tilde{\mathcal{F}}\mathrm{d}s.$$

The controller is designed in the same way as shown in (7.9)

$$u_i = -K\bar{\xi}_i - F\bar{\omega}_i = -[K, F]\bar{\chi}_i. \tag{7.32}$$

Under (7.32), the multi-agent systems (7.27) can be written as

$$\begin{aligned}\dot{\xi}_i &= A\xi_i - BK\bar{\xi}_i(t-\tau_u) - BF\bar{\omega}_i(t-\tau_u) + BF\omega_i + \tilde{f}_i(x_i, x_0) \\ &= (A - BK)\xi_i + B\bar{B}_1\bar{e}_i + \tilde{f}_i(x_i, x_0),\end{aligned}$$

where $\bar{B}_1 = [K, F]$.

Let $\xi = \left[\xi_1^T, \xi_2^T, \cdots, \xi_N^T\right]^T$ and $\tilde{f} = \left[\tilde{f}_1^T, \tilde{f}_2^T, \cdots, \tilde{f}_N^T\right]^T$. The closed-loop system is written as

$$\dot{\xi} = (I \otimes (A - BK))\xi + \left(I \otimes B\bar{B}_1\right)\bar{e} + \tilde{f}(x_i, x_0). \tag{7.33}$$

Then, the control and observer gains are chosen as

$$K = B^T\bar{P}_1, \qquad L = c\bar{P}_2^{-1}\bar{C}_\xi^T, \tag{7.34}$$

where $\bar{P}_1$ and $\bar{P}_2$ are positive definite matrices, c is the same constant as defined in (7.11).

For the stability analysis, first we need to establish bounds of the extra terms ϕ and $\bar{\lambda}$.

Lemma 7.3.1 *For the integral terms ϕ and $\bar{\lambda}$ shown in the error dynamics (7.31), bounds can be established as*

$$\|\phi\|^2 \le \tau \|\mathcal{L}_1\|_F^2 \bar{\rho}^2 e^{\bar{\alpha}\tau} \int_{t-\tau}^{t} \left(\gamma_1^2 \|\xi\|^2 + \gamma_2^2 \|\bar{e}_\omega\|^2\right) \mathrm{d}s,$$
$$\|\bar{\lambda}\|^2 \le \tau \|\mathcal{L}_1\|_F^4 \bar{\rho}^4 e^{2\bar{\alpha}\tau} \int_{t-\tau}^{t} \bar{e}^T(s-\tau)\bar{e}(s-\tau)\mathrm{d}s,$$

where $\bar{\alpha}$ is a positive number such that $\bar{\alpha} \ge \lambda_{\max}\left(A_\xi^T + A_\xi\right)$, and $\bar{\rho}$ is a positive number such that

$$\bar{\rho}^2 I \ge c^2 \bar{C}_z^T \bar{C}_z \bar{P}_2^{-1} \bar{P}_2^{-1} \bar{C}_z^T \bar{C}_z. \tag{7.35}$$

Proof 23 *Define* $\phi = \left[\phi_1^T, \phi_2^T, \cdots, \phi_N^T\right]^T$. *From (7.31), we have*

$$\phi_i = L\bar{C}_\xi \sum_{j=1}^{N} l_{ij} \int_{t-\tau}^{t} e^{A_\xi(t-s)} \tilde{\mathcal{F}}_j \mathrm{d}s = \sum_{j=1}^{N} l_{ij} \tilde{\phi}_j,$$

where $\tilde{\phi}_j = L\bar{C}_\xi \int_{t-\tau}^{t} e^{A_\xi(t-s)} \tilde{\mathcal{F}}_j \mathrm{d}s$. *Let* $\tilde{\phi} = \left[\tilde{\phi}_1^T, \tilde{\phi}_2^T, \cdots, \tilde{\phi}_N^T\right]^T$. *Then, we have*

$$\|\phi_i\| \leq \sum_{j=1}^{N} |l_{ij}| \left\|\tilde{\phi}_j\right\| \leq \|l_i\| \left\|\tilde{\phi}\right\|,$$

and

$$\|\phi\|^2 = \sum_{i=1}^{N} \|\phi_i\|^2 \leq \sum_{i=1}^{N} \|l_i\|^2 \left\|\tilde{\phi}\right\|^2 \leq \|\mathcal{L}_1\|_F^2 \left\|\tilde{\phi}\right\|^2.$$

Next we need to deal with $\left\|\tilde{\phi}\right\|^2$. *By Lemmas 2.1.1 and 2.1.2, we have*

$$\begin{aligned}
\left\|\tilde{\phi}_i\right\|^2 &= \int_{t-\tau}^{t} \tilde{\mathcal{F}}_i^T e^{A_\xi^T(t-s)} \mathrm{d}s \bar{C}_\xi^T L^T L \bar{C}_\xi \int_{t-\tau}^{t} e^{A_\xi(t-s)} \tilde{\mathcal{F}}_i \mathrm{d}s \\
&\leq \tau \bar{\rho}^2 \int_{t-\tau}^{t} \tilde{\mathcal{F}}_i^T e^{\left(A_\xi^T + A_\xi\right)(t-s)} \tilde{\mathcal{F}}_i \mathrm{d}s \\
&\leq \tau \bar{\rho}^2 e^{\bar{\alpha}\tau} \int_{t-\tau}^{t} \left\|\tilde{\mathcal{F}}_i\right\|^2 \mathrm{d}s \\
&\leq \tau \bar{\rho}^2 e^{\bar{\alpha}\tau} \int_{t-\tau}^{t} \left(\gamma_1^2 \|\xi_i\|^2 + \gamma_2^2 \|\bar{e}_{\omega i}\|^2\right) \mathrm{d}s.
\end{aligned}$$

Then, $\left\|\tilde{\phi}\right\|^2$ *can be bounded as*

$$\left\|\tilde{\phi}\right\|^2 = \sum_{i=1}^{N} \left\|\tilde{\phi}_i\right\|^2 \leq \tau \bar{\rho}^2 e^{\bar{\alpha}\tau} \int_{t-\tau}^{t} \left(\gamma_1^2 \|\xi\|^2 + \gamma_2^2 \|\bar{e}_\omega\|^2\right) \mathrm{d}s.$$

Therefore, we have

$$\|\phi\|^2 \leq \tau \|\mathcal{L}_1\|_F^2 \bar{\rho}^2 e^{\bar{\alpha}\tau} \int_{t-\tau}^{t} \left(\gamma_1^2 \|\xi\|^2 + \gamma_2^2 \|\bar{e}_\omega\|^2\right) \mathrm{d}s.$$

This completes the proof. The proof of $\bar{\lambda}$ *is similar to that of* λ *and hence omitted.*

Theorem 13 *For nonlinear multi-agent systems (7.25), the consensus disturbance rejection problem can be solved by the observer (7.29) and the controller (7.32) if there exists positive definite matrices $\bar{P}_1, \bar{P}_2$ and constants $\kappa_1, \kappa_2, \kappa_3, \rho > 0$, such that*

$$\begin{bmatrix} A\bar{W} + \bar{W}A^T - BB^T + \kappa_1 I & \bar{W} \\ \bar{W} & -\tilde{\gamma}^{-1} \end{bmatrix} < 0, \tag{7.36}$$

$$\bar{\rho}\bar{P}_2 - c\bar{C}_\xi^T \bar{C}_\xi \geq 0, \tag{7.37}$$

$$\begin{bmatrix} A_\xi^T \bar{P}_2 + \bar{P}_2 A_\xi - 2C_\xi^T C_\xi + \bar{H} & \bar{P}_2 \\ \bar{P}_2 & -\frac{1}{\kappa_2+\kappa_3} \end{bmatrix} < 0, \tag{7.38}$$

are satisfied with $\bar{W} = \bar{P}_1^{-1}$ and

$$\tilde{\gamma}_1 = \left(\kappa_1^{-1} + \bar{\gamma}_1\right)\gamma_1^2,$$
$$\bar{H} = g_{\min}^{-1} \bar{B}_1^T \bar{B}_1 + \bar{\sigma} I + g_{\min}^{-1} \bar{\gamma}_2 \bar{D},$$

where $\bar{\gamma}_1 = \kappa_2^{-1} g_{\max}\tau \left\|\mathcal{L}_1\right\|_F^2 \bar{\rho}^2 e^{\bar{\alpha}\tau}\gamma_1^2$, $\bar{\gamma}_2 = \kappa_2^{-1} g_{\max}\tau \left\|\mathcal{L}_1\right\|_F^2 \bar{\rho}^2 e^{\bar{\alpha}\tau}\gamma_2^2$, $\bar{\sigma} = e^\tau \kappa_3^{-1} g_{\max}\tau \left\|\mathcal{L}_1\right\|_F^4 \bar{\rho}^4 e^{2\bar{\alpha}\tau}$, $\bar{D} = [0_{n\times n}\ 0_{n\times s}; 0_{s\times n}\ I_{s\times s}]$.

Proof 24 *To start the consensus analysis, we try a Lyapunov function candidate*

$$\bar{V}_0 = \xi^T \left(I \otimes \bar{P}_1\right)\xi + \bar{e}^T \left(G \otimes \bar{P}_2\right)\bar{e}. \tag{7.39}$$

In view of (7.8) and (7.10), we have

$$\begin{aligned}
\dot{\bar{V}}_0 &= \xi^T \left[I \otimes \left(A^T\bar{P}_1 + \bar{P}_1 A - 2\bar{P}_1 BB^T\bar{P}_1\right)\right]\xi + 2\xi^T\left(I \otimes \bar{P}_1 B\bar{B}_1\right)\bar{e} \\
&\quad + \bar{e}^T\left[G \otimes \left(A_\xi^T\bar{P}_2 + \bar{P}_2 A_\xi\right) - c\left(G\mathcal{L}_1 + \mathcal{L}_1^T G\right)\otimes \bar{C}_\xi^T\bar{C}_\xi\right]\bar{e} \\
&\quad + 2\bar{e}^T\left(G\otimes\bar{P}_2\right)\bar{\lambda} + 2\xi^T\left(I\otimes\bar{P}_1\right)\tilde{f}\left(x_i, x_0\right) + 2\bar{e}^T\left(G\otimes\bar{P}_2\right)\phi \\
&\leq \xi^T\left[I\otimes\left(A^T\bar{P}_1 - \bar{P}_1 BB^T\bar{P}_1 + \kappa_1^{-1}\gamma_1^2 I + \kappa_1\bar{P}_1\bar{P}_1 + \bar{P}_1 A\right)\right]\xi \\
&\quad + \bar{e}^T\left[G\otimes\left(A_\xi^T\bar{P}_2 + \bar{P}_2 A_\xi + (\kappa_2+\kappa_3)\bar{P}_2\bar{P}_2 - 2\bar{C}_\xi^T\bar{C}_\xi\right) + \bar{B}_1^T\bar{B}_1\right]\bar{e} \\
&\quad + \kappa_2^{-1} g_{\max}\|\phi\|^2 + \kappa_3^{-1} g_{\max}\|\bar{\lambda}\|^2 \\
&\leq \xi^T\left[I\otimes\left(A^T\bar{P}_1 - \bar{P}_1 BB^T\bar{P}_1 + \kappa_1^{-1}\gamma_1^2 I + \kappa_1\bar{P}_1\bar{P}_1 + \bar{P}_1 A\right)\right]\xi \\
&\quad + \bar{e}^T\left[G\otimes\left(A_\xi^T\bar{P}_2 + \bar{P}_2 A_\xi + (\kappa_2+\kappa_3)\bar{P}_2\bar{P}_2 - 2\bar{C}_\xi^T\bar{C}_\xi\right) + \bar{B}_1^T\bar{B}_1\right]\bar{e} \\
&\quad + \kappa_2^{-1} g_{\max}\tau\left\|\mathcal{L}_1\right\|_F^2\bar{\rho}^2 e^{\bar{\alpha}\tau}\int_{t-\tau}^t\left(\gamma_1^2\|\xi\|^2 + \gamma_2^2\|\bar{e}_\omega\|^2\right)\mathrm{d}s \\
&\quad + \kappa_3^{-1} g_{\max}\tau\left\|\mathcal{L}_1\right\|_F^4\bar{\rho}^4 e^{2\bar{\alpha}\tau}\int_{t-\tau}^t \bar{e}^T(s-\tau)\bar{e}(s-\tau)\mathrm{d}s,
\end{aligned}$$

where Lemma 7.3.1 and Eq. (7.26) are used for the derivation.

For the integral terms, we consider the following Krasovskii functionals

$$\bar{V}_1 = \bar{\gamma}_1 \int_{t-\tau}^{t} \xi^T(s)\xi(s)\mathrm{d}s + \bar{\gamma}_2 \int_{t-\tau}^{t} \bar{e}_\omega^T(s)\bar{e}_\omega(s)\mathrm{d}s,$$
$$\bar{V}_2 = e^\tau \int_{t-\tau}^{t} \bar{e}^{s-t} e^T(s-\tau)\bar{e}(s-\tau)\mathrm{d}s + e^\tau \int_{t-\tau}^{t} \bar{e}^T(s)\bar{e}(s)\mathrm{d}s,$$

where $\bar{\gamma}_1 = \kappa_2^{-1} g_{\max}\tau \left\|\mathcal{L}_1\right\|_F^2 \bar{\rho}^2 e^{\bar{\alpha}\tau}\gamma_1^2$ *and* $\bar{\gamma}_2 = \kappa_2^{-1} g_{\max}\tau \left\|\mathcal{L}_1\right\|_F^2 \bar{\rho}^2 e^{\bar{\alpha}\tau}\gamma_2^2$ *are defined in Theorem 13.*

Let

$$V = \bar{V}_0 + \bar{V}_1 + \kappa_3^{-1} g_{\max}\tau \left\|\mathcal{L}_1\right\|_F^4 \bar{\rho}^4 e^{2\bar{\alpha}\tau}\bar{V}_2. \tag{7.40}$$

A direct evaluation gives that

$$\dot{V} \leq \xi^T \left(I \otimes \bar{H}_1\right)\xi + \bar{e}^T \left(G \otimes \bar{H}_2\right)\bar{e}, \tag{7.41}$$

where

$$\bar{H}_1 := A^T\bar{P}_1 + \bar{P}_1 A - \bar{P}_1 B B^T \bar{P}_1 + \kappa_1 \bar{P}_1\bar{P}_1 + \tilde{\gamma}_1 I, \tag{7.42}$$
$$\begin{aligned}\bar{H}_2 := {} & A_\xi^T \bar{P}_2 + \bar{P}_2 A_\xi + (\kappa_2+\kappa_3)\bar{P}_2\bar{P}_2 - 2C_\xi^T C_\xi \\ & + g_{\min}^{-1}\bar{B}_1^T\bar{B}_1 + \bar{\sigma} I + g_{\min}^{-1}\bar{\gamma}_2\bar{D},\end{aligned} \tag{7.43}$$

where $\tilde{\gamma}_1 = \left(\kappa_1^{-1} + \bar{\gamma}_1\right)\gamma_1^2$, $\bar{\sigma} = e^\tau\kappa_3^{-1} g_{\max}\tau \left\|\mathcal{L}_1\right\|_F^4 \bar{\rho}^4 e^{2\bar{\alpha}\tau}$, *and* $\bar{D} = [0_{n\times n}\ 0_{n\times s}; 0_{s\times n}\ I_{s\times s}]$ *are defined in Theorem 13.*

Following the same analysis in last section, the conditions $\bar{H}_1 < 0$ *and* $\bar{H}_2 < 0$ *are equivalent to the conditions specified in (7.36)–(7.38) which implies that* ξ *and* $\bar{e}$ *converge to zero asymptotically. Hence, the nonlinear consensus disturbance rejection is achieved.*

Remark 7.3.1 *Since the values of* τ, γ_1 and γ_2 *are fixed and they are not the decision variables of the LMIs, a feasible solution may not exist if these values are too large. To avoid this problem, a set of free parameters* κ_1, κ_2 and κ_3 *are introduced to provide more design degrees of freedom. In addition, if the delay* τ *is too large to find a feasible solution, an alternative way is to design a series of coupled predictors, each of which is responsible for prediction of one small portion of the delay [104].*

7.4 A Numerical Example

In this section, an example is used to demonstrate the potential applications of the proposed approach. Suppose a network of five unmanned aerial

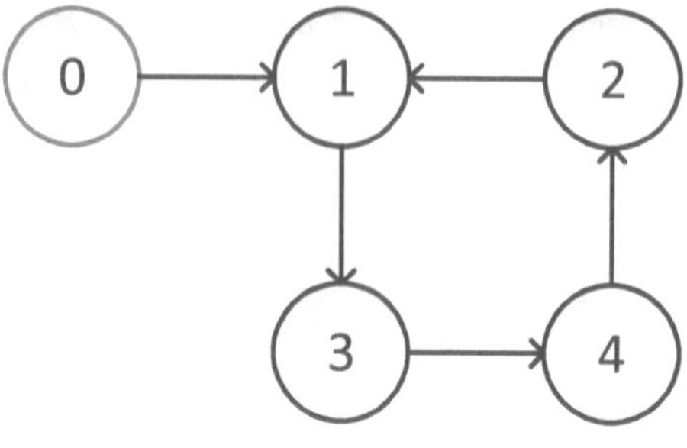

FIGURE 7.1: Communication topology.

vehicles are subject to the connection topology specified by the following adjacency matrix

$$\mathcal{A} = \begin{bmatrix} 0 & 0 & 0 & 0 & 0 \\ 1 & 0 & 1 & 0 & 0 \\ 0 & 0 & 0 & 0 & 1 \\ 0 & 1 & 0 & 0 & 0 \\ 0 & 0 & 0 & 1 & 0 \end{bmatrix}.$$

Note that the first row all are zeros, as the agent indexed by 0 is chosen as the leader. The communication graph in Figure 7.1 shows that only the follower indexed by 1 can get access to the leader and the communication topology contains a directed spanning tree.

The dynamics of the ith agent are described by (7.25), with

$$\dot{x}_i = \begin{bmatrix} x_{i1} \\ x_{i2} \end{bmatrix}, \ A = \begin{bmatrix} 0 & -1 \\ 1 & 0 \end{bmatrix}, \ B = \begin{bmatrix} 1 & 0.5 \\ 0.5 & 1 \end{bmatrix},$$
$$C = \begin{bmatrix} 1 & 0 \end{bmatrix}, f_1(x_i) = \beta_1 \begin{bmatrix} \sin(x_{i1}) & \sin(x_{i2}) \end{bmatrix}.$$

This practical dynamical model of UAV is given in [181]. In this scenario, it is supposed that only the output information is available. Our task in this example is to reject harmonic disturbances in the input channel which generated by a nonlinear external disturbance model (7.25) with

$$\dot{\omega}_i = \begin{bmatrix} \omega_{i1} \\ \omega_{i2} \end{bmatrix}, \ S = \begin{bmatrix} 0 & -0.1 \\ 0.1 & 0 \end{bmatrix}, F = \begin{bmatrix} 1 & 0 \\ 0 & 1 \end{bmatrix},$$
$$f_2 = \beta_2 \begin{bmatrix} (|\omega_{i1} + 1| - |\omega_{i1} - 1|) & 0 \end{bmatrix},$$

which represents an external periodic disturbance with known frequency but without any information of its magnitude and phase. The input delay $\tau_u = 0.05$s, the output delay $\tau_y = 0.05$s, and the Lipschitz constants are $\gamma_1 = \beta_1 = 0.03, \gamma_2 = 2\beta_2 = 0.04$. It can be checked that both Assumptions 7.1.1 and 7.1.2 are satisfied.

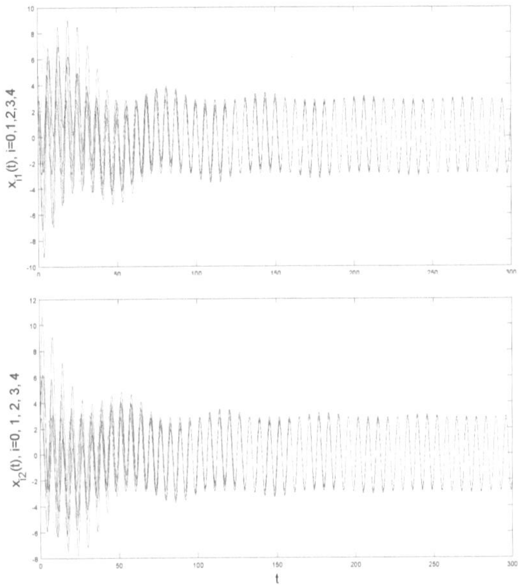

FIGURE 7.2: States trajectories of the five agents.

Following the result shown in Lemma 1.4.9, we obtain that $G = \text{diag}\{4,\ 5,\ 7,\ 6\}$ and $\rho_0 = 1.7995$. With $g_{\max} = 7$ and $2g_{\max}/\rho_0 = 7.7799$, we set $c = 8$ in (7.34).

The Laplacian matrix $\mathcal{L}_1$ associated with $\mathcal{A}$ is

$$\mathcal{L}_1 = \begin{bmatrix} 2 & -1 & 0 & 0 \\ 0 & 1 & 0 & -1 \\ -1 & 0 & 1 & 0 \\ 0 & 0 & -1 & 1 \end{bmatrix}.$$

The initial states of agents are chosen randomly within $[\ 0,\ 5\ ]$, and $u(\theta) = [\ 0,\ 0\ ]^T,\ \forall\theta \in [-\tau_u, 0], \bar{\chi}(\theta) = [\ 0,\ 0\ ]^T,\ \forall\theta \in [-\tau, 0]$. With $\bar{\rho} = 0.2, \kappa_1 = \kappa_2 = \kappa_3 = 0.1$, a feasible solution of the observer gain L is found to be $\begin{bmatrix} 0.8406 & -0.2526 & -0.1067 & -0.2220 \end{bmatrix}$, and a feedback gain K is found to be

$$K = \begin{bmatrix} 0.0364 & 0.0188 \\ 0.0182 & 0.0376 \end{bmatrix}.$$

Simulation study has been carried out with different disturbances for agents. Figure 7.2 shows the simulation results for the trajectories of the states. The control input and the relative output errors, $Y_i = \sum_{j=0}^{N} a_{ij}\,(y_i - y_j)\,,\ i = 1, 2, 3, 4$, are shown in Figures 7.3 and 7.4. The state and the disturbance observation errors are shown in Figures 7.5 and 7.6. From the results shown in these

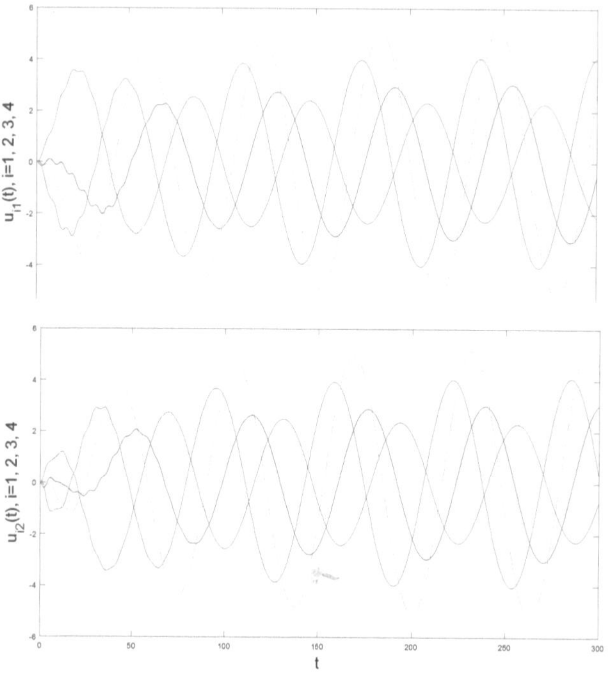

FIGURE 7.3: Control inputs of the four followers.

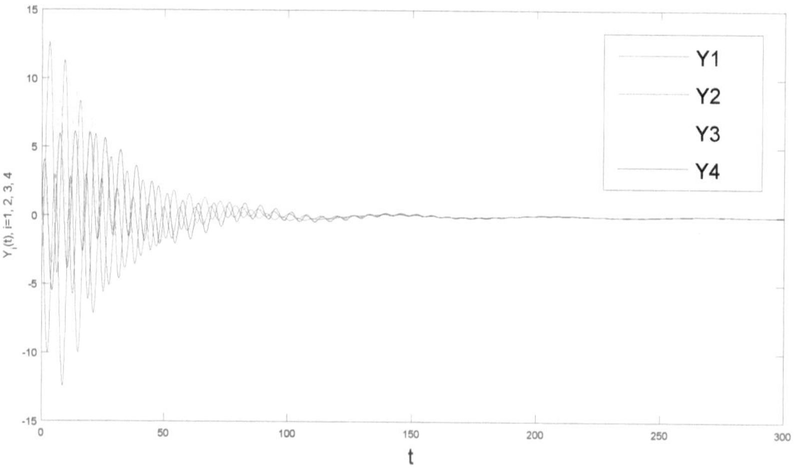

FIGURE 7.4: The relative output error of the four followers.

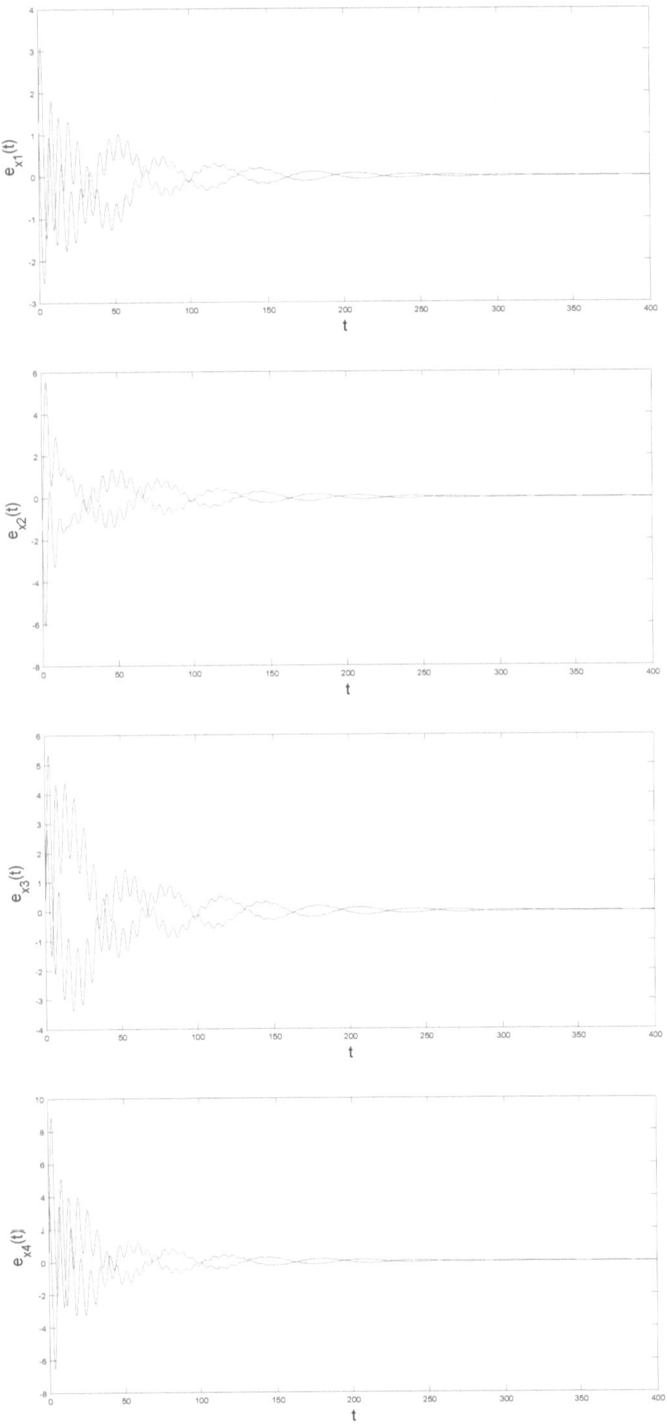

FIGURE 7.5: The estimation errors of the predictor-based state observers.

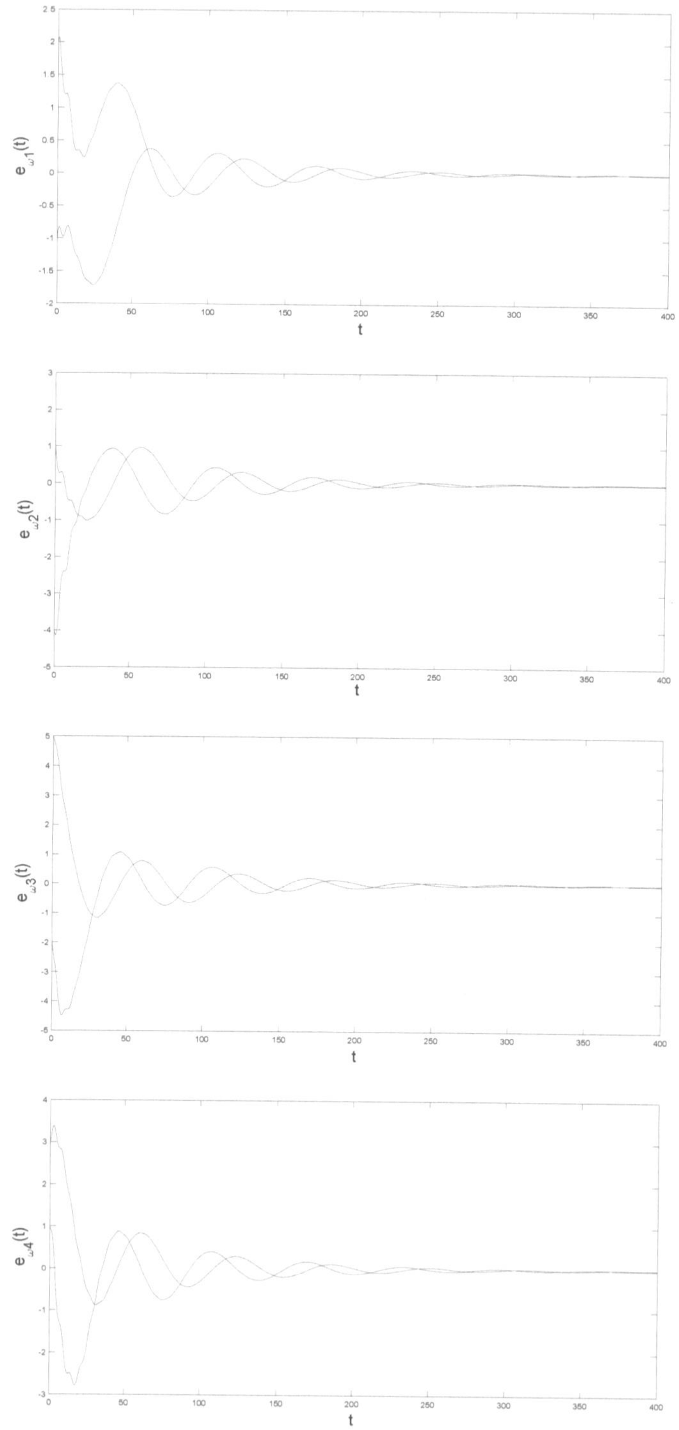

FIGURE 7.6: The estimation errors of the disturbance observers.

figures, it can be seen clearly that all the five agents reach consensus although they are under different disturbances. Therefore, the conditions specified in this chapter are sufficient to guarantee the consensus disturbance rejection. Furthermore, compared to the previous results in [164,210], the observer-based predictor controllers designed here have some good points. The first point is that the consensus disturbance rejection can be achieved via only relative output information. The second point is that this method not only can deal with input delay, but also can deal with output delay and nonlinear disturbances. Finally, the proposed observer-based predictor controllers are easier and safer to implement.

7.5 Conclusions

This chapter has investigated the observer-based output feedback consensus problem for multi-agent systems subject to external disturbance and delays in input and output. Novel predictor-based extended state observers are designed, and the delays and disturbances can be compensated efficiently with the proposed controllers. In particular, the observers and the state predictors do not contain any integral term of the past control input, which greatly reduces the computation burden and improves the practical implementation. Consensus analysis is put into the framework of Lyapunov-Krasovskii functionals and sufficient conditions are derived to guarantee that the consensus errors converge to zero asymptotically. The results have been also extended to nonlinear multi-agent systems with nonlinear disturbances. Simulation results show the validity of the proposed method and design.

7.6 Notes

This chapter extends the ideas in [104, 164, 201] and develops consensus protocols for leader-follower nonlinear multi-agent systems with directed topology. The main materials of this chapter are based on [163]. It is worth mentioning that the control protocols proposed in this chapter are not fully distributed. The results in this chapter can be extended to the fully distributed cases by following the procedures shown in [169].

Chapter 8

Formation Control with Disturbance Rejection for a Class of Lipschitz Nonlinear Systems

This chapter addresses the leader-follower formation control problem for general multi-agent systems with Lipschitz nonlinearity and unknown disturbances. To deal with the disturbances, a disturbance observer-based control (DOBC) strategy is developed for each follower. Then, a time-varying formation protocol is proposed based on the relative state of the neighbouring agents and sufficient conditions for global stability of the formation control are identified using Lyapunov method in the time domain. The proposed strategy and analysis guarantee that all signals in the closed-loop dynamics are uniformly ultimately bounded and the formation tracking error converges to an arbitrarily small residual set. Finally, the validity of the proposed controller is demonstrated through a numerical example.

8.1 Problem Formulation

8.1.1 Problem Statement

Consider a group of $N+1$ agents, consisting of N followers and one leader indexed by 0

$$\dot{x}_i(t) = Ax_i(t) + Bu_i(t) + \phi(x_i(t)) + BF\omega_i(t), \tag{8.1}$$

where for agent i, $i = 0, 1, \ldots, N$, $x_i \in \mathbb{R}^n$ is the state, $u_i \in \mathbb{R}^m$ is the control input, $A \in \mathbb{R}^{n\times n}$, $B \in \mathbb{R}^{n\times m}$ and $F \in \mathbb{R}^{m\times s}$ are constant matrices with (A, B) being controllable, the unknown nonlinear function $\phi : \mathbb{R}^n \to \mathbb{R}^n$, $\phi(0) = 0$, is assumed to satisfy the Lipschitz condition as

$$\|\phi(\alpha) - \phi(\beta)\| \leq \gamma\|\alpha - \beta\|, \forall \alpha, \beta \in \mathbb{R}^n,$$

where $\gamma > 0$ is the Lipschitz constant, and $\omega_i \in \mathbb{R}^s$ is a disturbance that is generated by a linear exogenous system

$$\dot{\omega}_i(t) = S\omega_i(t), \tag{8.2}$$

with $S \in \mathbb{R}^{s\times s}$ is known and (S, BF) is observable. For the leader-follower structure, it is reasonable to assume that the leader has no neighbours, and the leader's control input is zero, i.e. $u_0 = 0,\ \omega_0 = 0$.

The objective of this chapter is to design a distributed formation protocol for each follower, such that the formation tracking errors $e_i(t) = x_i(t) - d_i(t) - x_0(t)$ are uniformly ultimately bounded for any initial condition $x_i(0)$, $i = 0, 1, \cdots, N$, where $d_i(t)$ is a specified time-varying formation pattern between the leader and the ith follower. For the convenience, we define $d_0(t) \equiv 0,\ e_0(t) \equiv 0$.

Assumption 8.1.1 *The eigenvalues of S are distinct, and lie on the imaginary axis.*

Assumption 8.1.2 *The specified formation vector $d(t) = \left[d_1^T, d_1^T, \cdots, d_N^T\right]^T$ is bounded with $d_i(t),\ \forall\ i = 1,\ 2,\ \cdots,\ N$ continuously differentiable, i.e., $\|d(t)\| \le \epsilon_0$, where ϵ_0 is a positive constant.*

Assumption 8.1.3 *The ith agent can obtains its neighbors' formation information via the inter-agent communication.*

Assumption 8.1.4 *The communication topology $\mathcal{G}$ contains a directed spanning tree with the leader as the root.*

With Assumption 8.1.4, we know that the Laplacian matrix $\mathcal{L}$ of the communication topology $\mathcal{G}$ has simple zero eigenvalue with right eigenvector $\mathbf{1} = [1, 1, \ldots, 1]^T$ and all the other eigenvalues of $\mathcal{L}$ have positive real parts [76]. Since the leader has no neighbours, the Laplacian matrix $\mathcal{L}$ has the following structure

$$\mathcal{L} = \begin{bmatrix} 0 & 0_{1\times N} \\ \mathcal{L}_2 & \mathcal{L}_1 \end{bmatrix},$$

where $\mathcal{L}_1 \in \mathbb{R}^{N\times N}$ and $\mathcal{L}_2 \in \mathbb{R}^{N\times 1}$. It can be seen that $\mathcal{L}_1$ is a nonsingular M-matrix and there exists a positive diagonal matrix Q such that

$$Q\mathcal{L}_1 + \mathcal{L}_1^T Q \ge \rho_0 I, \tag{8.3}$$

for some positive constant ρ_0. Q can be constructed by letting $Q = \mathrm{diag}\{q_1, q_2, \cdots, q_N\}$, where $q = [q_1, q_2, \cdots, q_N]^T = \left(\mathcal{L}_1^T\right)^{-1} [1, 1, \cdots, 1]^T$.

Lemma 8.1.1 *[78] For a given continuous system $\dot{z} = y(z, t)$, $y(\cdot)$ is assumed locally Lipschitz in z. If there exists a differentiable function $V(z, t) \ge 0$ such that*

$$\beta_1(\|z\|) \le V(z, t) \le \beta_2(\|z\|),$$
$$\dot{V}(z, t) \le -\beta_3(\|z\|) + \iota,$$

where $\iota > 0$ is a constant, β_1, β_2 belong to class $\mathcal{K}_\infty$ functions, and β_3 belongs to class $\mathcal{K}$ function. The solution $z(t)$ of the system $\dot{z} = y(z, t)$ is uniformly ultimately bounded.

8.2 DOBC-Based Formation Control

With the formation tracking errors $e_i(t) = x_i(t) - d_i(t) - x_0(t)$, we have

$$\dot{e}_i(t) = Ae_i(t) + Bu_i(t) + \varepsilon_i(t) + BF\omega_i(t) + Ad_i(t) - \dot{d}_i(t), \tag{8.4}$$

where $\varepsilon_i(t) = \phi(x_i) - \phi(x_0)$. Define $e(t) = [e_1^T(t), e_2^T(t), \cdots, e_N^T(t)]^T$. Then, the leader-follower formation of system (8.1) is achieved when $\lim_{t\to\infty} e(t) = 0$, as $e = 0$ implies that $x_0 = x_1 - d_1 = \cdots = x_N - d_N$. Therefore, the formation problem of system (8.1) is transformed into the regulator problem of system (8.4).

With the relative state of the neighbouring agent

$$\begin{aligned} z_i(t) &= \sum_{j=0}^{N} a_{ij} \left[(x_i(t) - d_i(t)) - (x_j(t) - d_j(t))\right] \\ &= \sum_{j=0}^{N} a_{ij} \left[e_i(t) - e_j(t)\right] \\ &= \sum_{j=1}^{N} a_{ij} \left[e_i(t) - e_j(t)\right] + a_{i0} e_i(t) \\ &= \sum_{j=1}^{N} l_{ij} e_j(t), \end{aligned} \tag{8.5}$$

a formation controller is proposed as

$$u_i(t) = cKz_i + v_i(t) - F\hat{\omega}_i,$$

where $c \geq 2q_{\max}/\rho_0$ is a positive real constant with $q_{\max} = \max\{q_1, q_2, \cdots, q_N\}$, $v_i(t) \in \mathbb{R}^m$ is the external command input, $\hat{\omega}_i(t)$ is generated by the following disturbance observer

$$\begin{cases} \hat{\omega}_i = \eta_i + Lz_i, \\ \dot{\eta}_i = S\eta_i + (SL - LA)\, z_i - cLBK \sum_{j=1}^{N} l_{ij} z_j, \end{cases} \tag{8.6}$$

where $K \in \mathbb{R}^{m\times n}$ and $L \in \mathbb{R}^{s\times n}$ are the control and the observer gain to be designed later.

By (8.5), the closed-loop error dynamics of the ith agent can be obtained as

$$\dot{e}_i(t) = Ae_i(t) + cBK \sum_{j=1}^{N} l_{ij} e_j(t) + \varepsilon_i(t) - BF\tilde{\omega}_i(t) + Bv_i(t) + Ad_i(t) - \dot{d}_i(t), \tag{8.7}$$

where $\tilde{\omega}_i(t) = \hat{\omega}_i - \omega_i$. Therefore, the formation of systems (8.1) is achieved if the following systems

$$\dot{e}_i(t) = Ae_i(t) + cBK\sum_{j=1}^{N} l_{ij}e_j(t) + \varepsilon_i(t) - BF\tilde{\omega}_i(t), \tag{8.8}$$

are asymptotically stable, and

$$\lim_{t\to\infty}\left(Bv_i(t) + Ad_i(t) - \dot{d}_i(t)\right) = 0, \ \forall\, i = 1, 2, \cdots, N. \tag{8.9}$$

With the controller and observer shown in (8.5) and (8.6), K and L are chosen as

$$K = -B^T P, \tag{8.10}$$

$$L = \rho_1 M^{-1}(BF)^T, \tag{8.11}$$

where $P > 0,\ M > 0$ are constant matrices to be designed, ρ_1 is a scaler.

Furthermore, the nonlinear term $\varepsilon(x)$ in the new system (8.7) is related to state $x(t)$. For the formation stability analysis, a bound of this term needs to be found in terms of the transformed state $e(t)$.

Lemma 8.2.1 *For the nonlinear function $\varepsilon(x)$ in (8.7), a bound can be established in terms of the state $e(t)$ as*

$$\|\varepsilon\|^2 \le 2\gamma^2(\|e(t)\|^2 + \|d(t)\|^2). \tag{8.12}$$

Proof 25 *Based on the state transformation $e_i(t) = x_i(t) - d_i(t) - x_0(t)$, and $\varepsilon_i = \phi(x_i) - \phi(x_0)$, we have*

$$\|\varepsilon_i\| = \|\phi(x_i) - \phi(x_0)\| \le \gamma\,\|x_i - x_0\|\,. \tag{8.13}$$

It then follows that

$$\|\varepsilon_i\| \le \gamma\,\|e_i(t) + d_i(t)\| \le \gamma\,(\|e_i(t)\| + \|d_i(t)\|)\,, \tag{8.14}$$

and

$$\|\varepsilon\|^2 = \sum_{i=1}^{N}\|\varepsilon_i\|^2 \le 2\gamma^2\sum_{i=1}^{N}\left(\|e_i(t)\|^2 + \|d_i(t)\|^2\right) = 2\gamma^2\left(\|e(t)\|^2 + \|d(t)\|^2\right), \tag{8.15}$$

where we have used $\sum_{k=1}^{N}\|e_i\|^2 = \|e\|^2$ and the inequality $(a+b)^2 \le 2(a^2+b^2)$. This completes the proof.

Theorem 14 *For systems (8.8), the robust stability problem with disturbance rejection can be solved by the control algorithm (8.5) with (8.10)-(8.11) if there exist positive definite matrices* P, M *and constants* $\rho_1, \kappa > 0$, *such that*

$$\begin{bmatrix} AW + WA^T - 2BB^T + 2\kappa I & W \\ W & -\frac{q_{\min}}{2\gamma^2 r_1 + \kappa_1} \end{bmatrix} < 0, \tag{8.16}$$

$$MS + S^T M - r_2 F^T B^T BF + r_3 I < 0, \tag{8.17}$$

are satisfied with $W = P^{-1}$, $r_1 = q_{\max}(\rho_1 \sigma_{\max}^2(\mathcal{L}_1) + 1)/\kappa$, $r_2 = 2\rho_1/c - q_{\max}/\kappa q_{\min}^{-1}$, $q_{\min} = \min\{q_1, q_2, \cdots, q_N\}$, $q_{\max} = \max\{q_1, q_2, \cdots, q_N\}$, *and* κ_1 *is any positive number.*

Proof 26 *Systems (8.8) can be written in the compact form as*

$$\dot{e}(t) = (I_N \otimes A)e(t) + c(\mathcal{L}_1 \otimes BK)e(t) + \varepsilon(t) - (I_N \otimes BF)\tilde{\omega}(t), \tag{8.18}$$

where $\varepsilon = \left[\varepsilon_1^T, \varepsilon_2^T, \cdots, \varepsilon_N^T\right]^T, \tilde{\omega} = \left[\tilde{\omega}_1^T, \tilde{\omega}_2^T, \cdots, \tilde{\omega}_N^T\right]^T$. *For disturbance observer error* $\tilde{\omega}_i(t)$, *with condition (8.9), a direct differentiation gives that*

$$\begin{aligned} \dot{\tilde{\omega}}_i(t) &= \dot{\eta}_i(t) + L\sum_{j=1}^{N} l_{ij}\dot{e}_j(t) - S\omega_i(t) \\ &= S\tilde{\omega}_i(t) - LBF\sum_{j=1}^{N} l_{ij}\tilde{\omega}_j(t) + L\sum_{j=1}^{N} l_{ij}\varepsilon_j, \end{aligned} \tag{8.19}$$

which can be written in the compact form as

$$\dot{\tilde{\omega}}(t) = (I_N \otimes S)\tilde{\omega}(t) - (\mathcal{L}_1 \otimes LBF)\tilde{\omega}(t) + (\mathcal{L}_1 \otimes L)\varepsilon. \tag{8.20}$$

Consider the Lyapunov function candidate

$$V(t) = e^T(t)\,(Q \otimes P)\,e(t) + \tilde{\omega}^T(t)\,(Q \otimes M)\,\tilde{\omega}(t). \tag{8.21}$$

The derivative of $V(t)$ *along the trajectory of (8.18) and (8.19) can be obtained as*

$$\begin{aligned} \dot{V}(t) = {} & e^T(t)\left(Q \otimes \left(PA + A^T P\right) - c\left(Q\mathcal{L}_1 + \mathcal{L}_1^T Q\right) \otimes PBB^T P\right) e(t) \\ & + \tilde{\omega}^T(t)\left(Q \otimes \left(MS + S^T M\right) - \rho_1(Q\mathcal{L}_1 + \mathcal{L}_1^T Q) \otimes F^T B^T BF\right)\tilde{\omega}(t) \\ & + 2\sum_{i=2}^{N} q_i e_i^T(t) P\left(\varepsilon_i - BF\tilde{\omega}_i(t)\right) - 2\rho_1\tilde{\omega}^T(t)(Q\mathcal{L}_1 \otimes (BF)^T)\varepsilon \\ \leq {} & e^T(t)\left(Q \otimes \left(PA + A^T P\right) - c\rho_0 I \otimes PBB^T P\right) e(t) \\ & + \tilde{\omega}^T(t)\left(Q \otimes \left(MS + S^T M\right) - \rho_1\rho_0 I \otimes F^T B^T BF\right)\tilde{\omega}(t) \\ & + 2\sum_{i=2}^{N} q_i e_i^T(t) P\left(\varepsilon_i - BF\tilde{\omega}_i(t)\right) - 2\rho_1\tilde{\omega}^T(t)(Q\mathcal{L}_1 \otimes (BF)^T)\varepsilon \end{aligned}$$

$$\begin{aligned}
&\leq e^T(t)\left(Q\otimes\left(PA+A^TP+2\kappa PP\right)-c\rho_0 I\otimes PBB^TP\right)e(t)\\
&\quad+\frac{q_{\max}}{\kappa}\|\varepsilon\|^2-2\rho_1\tilde{\omega}^T(t)(Q\mathcal{L}_1\otimes(BF)^T)\varepsilon\\
&\quad+\tilde{\omega}^T\left(Q\otimes\left(MS+S^TM+\frac{q_{\max}}{\kappa q_{\min}}F^TB^TBF\right)-\rho_1\rho_0 I\otimes F^TB^TBF\right)\tilde{\omega}\\
&\leq e^T(t)\left(Q\otimes\left(PA+A^TP+2\kappa PP-2PBB^TP\right)\right)e(t)+r_1\|\varepsilon\|^2\\
&\quad+\tilde{\omega}^T(t)\left(Q\otimes\left(MS+S^TM-r_2F^TB^TBF+r_3I\right)\right)\tilde{\omega}(t),
\end{aligned}\tag{8.22}$$

where $r_1 = q_{\max}(\rho_1\sigma_{\max}^2(\mathcal{L}_1)+1)/\kappa$, $r_2 = 2\rho_1/c - q_{\max}/\kappa q_{\min}^{-1}$, $r_3 = q_{\max}\rho_1\kappa\lambda_{\max}(F^TB^TBF)/q_{\min}$.

From (8.22) and (8.12), we obtain that

$$\begin{aligned}
\dot{V}(t)&\leq e^T(t)\left(Q\otimes\left(PA+A^TP+2\kappa PP-2PBB^TP+2\gamma^2r_1/q_{\min}I\right)\right)e(t)\\
&\quad+\tilde{\omega}^T(t)\left(Q\otimes\left(MS+S^TM-r_2F^TB^TBF+r_3I\right)\right)\tilde{\omega}(t)+2\gamma^2r_1\|d(t)\|^2\\
&\leq e^T(t)\left(Q\otimes P_1\right)e(t)+\tilde{\omega}^T(t)\left(Q\otimes M_1\right)\tilde{\omega}(t)+2\gamma^2r_1\|d(t)\|^2,
\end{aligned}\tag{8.23}$$

where

$$P_1 = PA+A^TP+2\kappa PP-2PBB^TP+2\gamma^2r_1/q_{\min}I,\tag{8.24}$$

$$M_1 = MS+S^TM-r_2F^TB^TBF+r_3I.\tag{8.25}$$

With (8.24) and (8.25), it can be shown by Schur Complement that conditions (8.16) and (8.17) are respectively equivalent to $P_1 < -\kappa_1/q_{\min}I$ *and* $M_1 < -\kappa_1/q_{\min}I$, *which further implies from (8.23) that* $\dot{V}(t) < -\kappa_1\left(\|e(t)\|^2+\|\tilde{\omega}(t)\|^2\right)+2\gamma^2r_1\|d(t)\|^2$. *Since* $d(t)$ *is bounded as mentioned in Assumption 8.1.2, the positive term could be very small by choosing appropriate free scaler* κ. *By Lemma 8.1.1, we have that the tracking error* $e(t)$ *is uniformly ultimately bounded and the disturbance rejection is achieved.*

Remark 8.2.1 *The state transformation can only apply to the linear parts, and the nonlinear functions remain functions of the original state, which leads to extra complexity in the stability analysis. Due to the nonlinear terms, the formation tracking errors can only be uniformly ultimately bounded instead of converging to zero. If the Lipschitz constant* $\gamma \equiv 0$, *the formation tracking errors will converge to zero and the systems (8.8) will asymptotically stable at the origin. Besides, for the disturbance observer (8.6), the information of neighbors' neighbors is required due to the non-identical disturbances with unknown amplitudes and phases.*

Remark 8.2.2 *The formation feasibility condition (8.9) can be checked in a way similar to the linear mult-agent systems counterpart in [32]. Let* $\hat{B} = [\bar{B}^T, \tilde{B}^T]^T$ *be a nonsingular matrix with* $\bar{B} \in \mathbb{R}^{m\times n}$ *and* $\tilde{B} \in \mathbb{R}^{(n-m)\times n}$, *such that* $\bar{B}B = I$ *and* $\tilde{B}B = 0$. *First step is to check the feasibility of the following condition*

$$\lim_{t\to\infty}\left(\tilde{B}Ad_i(t)-\tilde{B}\dot{d}_i(t)\right)=0,\ \forall\, i=1,2,\cdots,N.\tag{8.26}$$

If (8.26) is satisfied, then the external command input $v_i(t)$ is chosen as $v_i(t) = -\bar{B}Ad_i(t) + \bar{B}\dot{d}_i(t)$ to guarantee that

$$\lim_{t\to\infty} \left(\bar{B}Ad_i(t) + v_i(t) - \bar{B}\dot{d}_i(t)\right) = 0,\ \forall\ i = 1, 2, \cdots, N. \tag{8.27}$$

With (8.26) and (8.27), the condition (8.9) is guaranteed. The proof is given by Theorem 3 in [32].

Remark 8.2.3 *Note that $q_{\min}$, $q_{\max}$, r_1, r_2 and r_3 can be easily calculated from the Laplacian matrix of any given network connection. It is worth mentioning that the Laplacian matrix $\mathcal{L}$ is a global information. In this sense, the formation control protocols proposed in this chapter are not fully distributed. The results in this chapter can be extended to the fully distributed cases by following the procedures shown in [28].*

8.3 A Numerical Example

In this section, we will demonstrate the formation control method with disturbance rejection under the leader-follower setup of five subsystems subject to the connection topology specified by the following adjacency matrix

$$\mathcal{A} = \begin{bmatrix} 0 & 0 & 0 & 0 & 0 \\ 1 & 0 & 0 & 1 & 0 \\ 0 & 1 & 0 & 0 & 0 \\ 0 & 0 & 1 & 0 & 1 \\ 1 & 1 & 0 & 0 & 0 \end{bmatrix}.$$

The communication graph is shown in Figure 8.1, from which it shows that only the followers indexed by 1 and 4 can get access to the leader and the communication topology contains a directed spanning tree. The dynamics of the ith agent are described by (8.1), with

$$A = \begin{bmatrix} -1 & 1 \\ 0 & 0 \end{bmatrix},\ B = \begin{bmatrix} 0 \\ 1 \end{bmatrix}, \phi(x_i) = g \begin{bmatrix} \sin(x_{i1}(t)) \\ \sin(x_{i2}(t)) \end{bmatrix}.$$

In this scenario, it is supposed that external disturbances exist in the control channel. The external disturbance $w_i(t)$ is generated by (8.2) with

$$S = \begin{bmatrix} 0 & -0.1 \\ 0.1 & 0 \end{bmatrix},\quad F = \begin{bmatrix} 1 & 1 \end{bmatrix},$$

which represents an external periodic disturbance with known frequency but without any information of its magnitude and phase. The Lipschitz constant

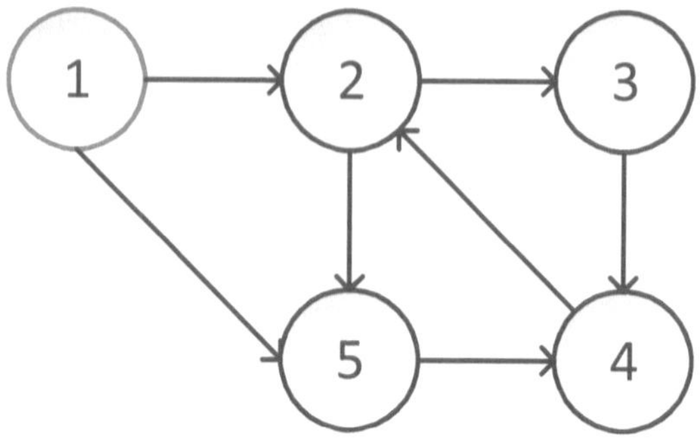

FIGURE 8.1: Communication topology.

is $\gamma = g = 0.03$. It can be checked that both Assumptions 8.1.1 and 8.1.4 are satisfied. The Laplacian matrix $\mathcal{L}_1$ associated with $\mathcal{A}$ is that

$$\mathcal{L}_1 = \begin{bmatrix} 2 & 0 & -1 & 0 \\ -1 & 1 & 0 & 0 \\ 0 & -1 & 2 & -1 \\ -1 & 0 & 0 & 2 \end{bmatrix}.$$

Following (8.3), we obtain that $Q = \text{diag}\{0.3846 \;\; 0.3571 \;\; 0.5556 \;\; 0.7143\}$ and $\rho_0 = 0.2573$. With $q_{\max} = 0.7143$ and $2q_{\max}/\rho_0 = 5.5523$, we set $c = 6$ in the control input (8.5).

The initial states of agents are chosen randomly within $[-20, \; 20]$. With the conditions (8.16) and (8.17), feasible solutions of the feedback gain K and the observer gain L are found to be

$$K = \begin{bmatrix} 0.0333 & -0.6113 \end{bmatrix}, \quad L = \begin{bmatrix} 0 & 21.6926 \\ 0 & 1.6114 \end{bmatrix}.$$

The formation is defined as follows:

$$d_i(t) = \begin{bmatrix} \sin(t + \frac{(i-1)\pi}{2}) - \cos(t + \frac{(i-1)\pi}{2}) \\ 2\sin(t + \frac{(i-1)\pi}{2}) \end{bmatrix},$$

which presents a periodic time-varying formation and keep rotating around the leader. Simulation study has been carried out with different disturbances for the followers. The trajectories of five agents are shown as Figure 8.2. Figure 8.3 shows the trajectory snapshots with $t = 0s, \; 5s, \; 10s, \; 15s, \; 20s$. It can be seen that the time-varying formation is achieved with predefined formation reference. The formation tracking errors between the four followers and the leader are shown in Figures 8.4 and 8.5. The disturbance observation errors are shown in Figure 8.6. From the results shown in these figures, we can see that all the five agents reach formation although they are subject to different disturbances. Therefore, the conditions specified in Theorem 14 and (8.26)–(8.27) are sufficient to guarantee the formation disturbance rejection.

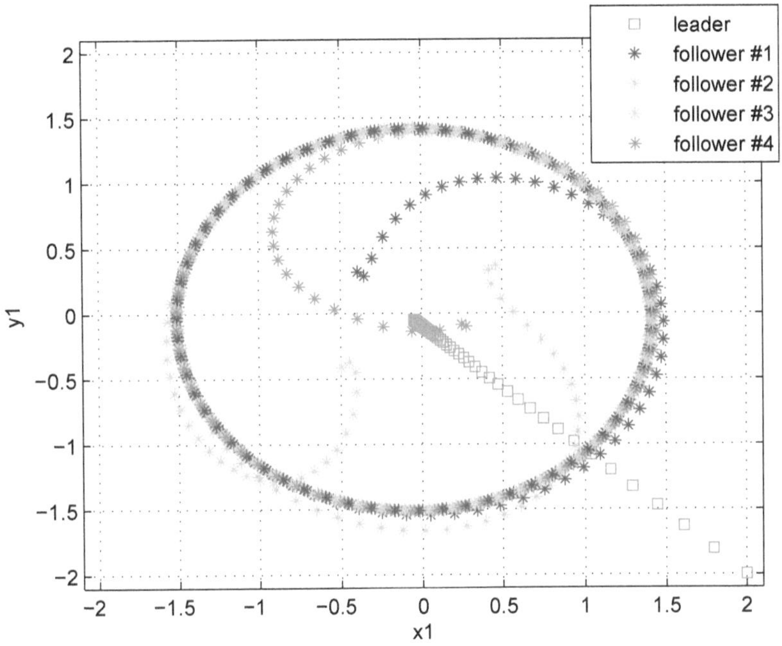

FIGURE 8.2: Trajectories of the leader and followers.

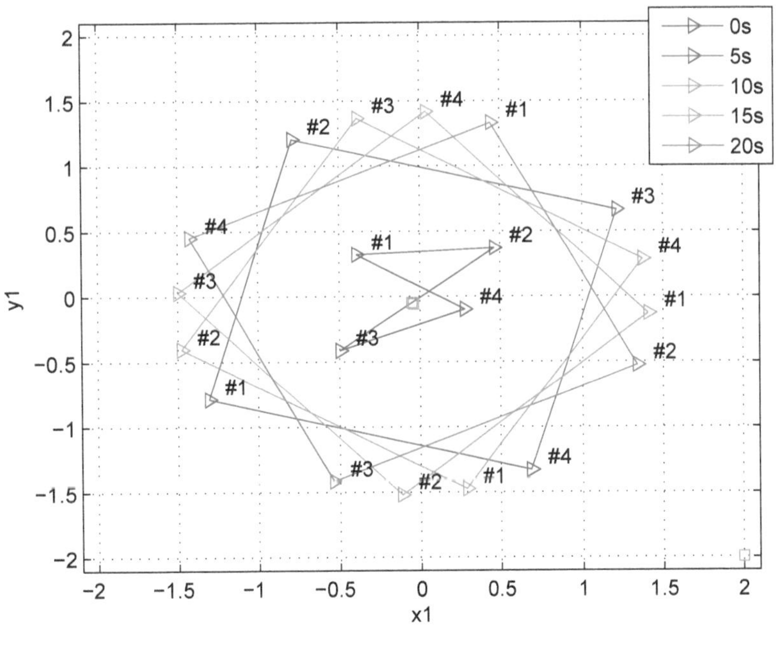

FIGURE 8.3: Trajectory snapshots at time $t = 0s,\ 5s,\ 10s,\ 15s,\ 20s$.

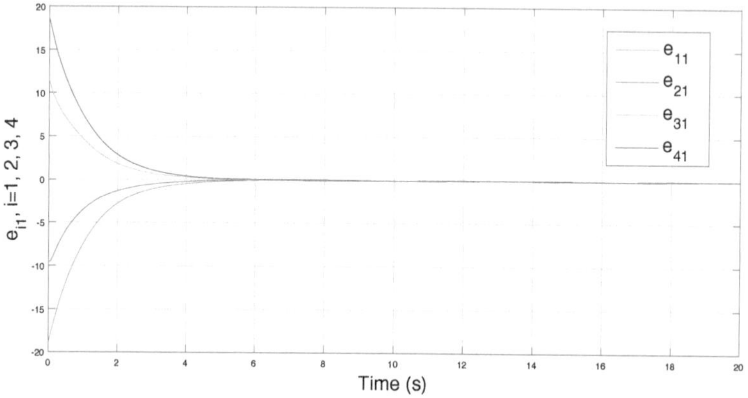

FIGURE 8.4: The formation error of state 1 with $g = 0.05$.

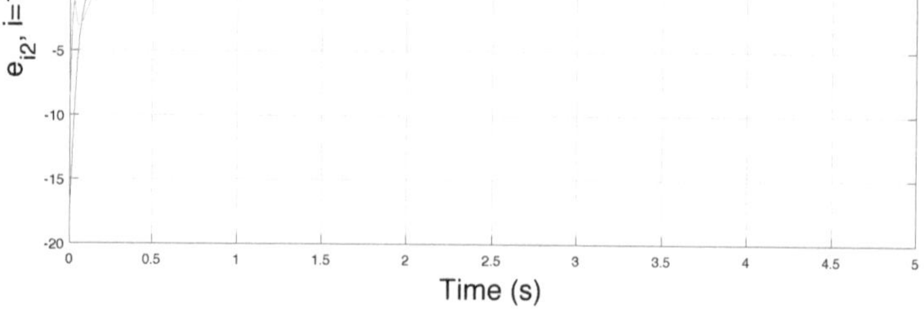

FIGURE 8.5: The formation error of state 2 with $g = 0.05$.

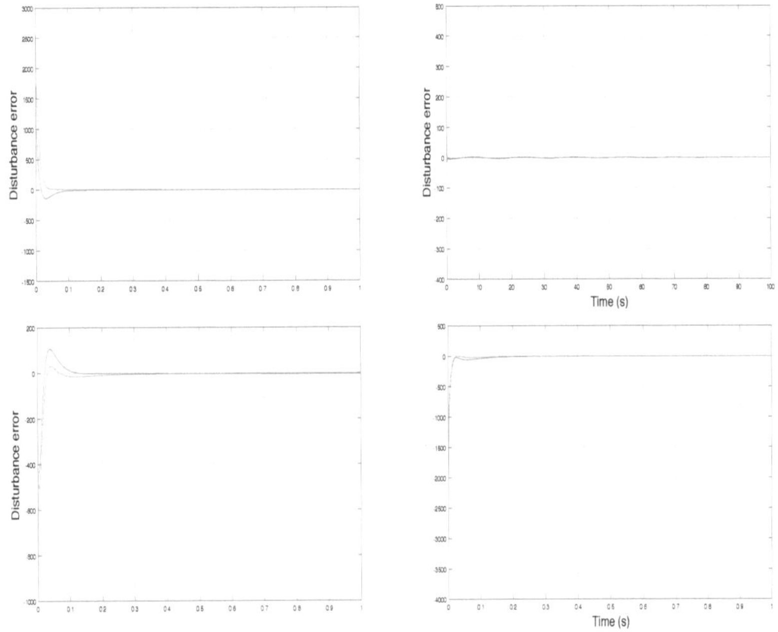

FIGURE 8.6: The estimation errors of the disturbance observers.

8.4 Conclusions

In this chapter, the formation control problem for a class of Lipschitz nonlinear multi-agent systems with external disturbances has been addressed. Disturbance observer and time-varying formation protocol is proposed for each follower based on the relative state and formation information of the neighbouring agents. The influence of the Lipschitz nonlinearity has been taken into account in the formation stability analysis and sufficient conditions for global stability of the formation control are identified using Lyapunov method in the time domain. The proposed strategy and analysis guarantee that all signals in the closed-loop dynamics are uniformly ultimately bounded and the formation tracking errors converge to an arbitrarily small residual set. Finally, the effectiveness of the theoretical results has been illustrated by a numerical example.

8.5 Notes

This chapter extends an idea in [28,158] and develops disturbance rejection formation protocols for Lipschitz nonlinear multi-agent networks with directed information flows. The main materials of this chapter are based on [160]. It is worth mentioning that the formation control protocols proposed in this chapter are not fully distributed. The results in this chapter can be extended to the fully distributed cases by following the procedures shown in [76,169].

Chapter 9

Fixed-Time Formation Control of Input-Delayed Multi-Agent Systems: Design and Experiments

Time delays exist in network-connected systems. Especially for vision-based multi-robot systems, time delays are diverse and complicated due to the communication network, camera latency, image processing, etc. At the same time, many tasks, such as searching and rescue, have timing requirement. This chapter focuses on fixed-time formation control of multi-robot systems subject to delay constraints. First, predictor-based state transformation is employed for each robot to deal with the input delay and the uncertain terms remained in the transformed systems are carefully considered. Then, a couple of nonlinear fixed-time formation protocols are proposed for the multi-robot systems with undirected and directed topologies, and the corresponding settling time is derived by using the Lyapunov functions. In particular, the upper-bound estimation of the formation settling time is explicitly given irrelevant to the initial conditions. Finally, the protocols are validated through a numerical simulation example and then implemented on an E-puck robots platform. Both simulation and experimental results demonstrate the effectiveness of the proposed formation protocols.

9.1 Problem Formulation and Preliminaries

9.1.1 Problem Formulation

Consider a group of N mobile robots with single-integrator dynamics:

$$\dot{x}_i(t) = u_i(t - h_i), \tag{9.1}$$

where for agent i, $i = 1, 2, \ldots, N$, $x_i \in \mathbb{R}^p$ is the state, $u_i \in \mathbb{R}^p$ is the control input, and h_i is the delay time. For the convenience of presentation, in this chapter, it is assumed that $p = 1$. In practice, the delay time for each robot may be slightly different due to disturbances or unstable connection. Thus, we define h_i in a more standard form $h_i = h + \Delta h_i$.

FIGURE 9.1: E-puck robot.

Assumption 9.1.1 *h is a known constant and Δh_i is unknown and bounded by $\|\Delta h_i\| \leq R_1$.*

Remark 9.1.1 *In this chapter, we consider that the delay consists of a nominal part h perturbed by a non-identical uncertain part Δh_i which can be constant or time-varying. We will use the robust control technique to deal with this uncertainty. The only requirement for Δh_i is that it is bounded by a known constant. However, the distribution of Δh_i is not needed to be known.*

Assumption 9.1.2 *The second derivative of the state $x_i(t)$ is bounded, i.e., $|\ddot{x}_i(t)| \leq R_2$, where R_2 is a constant.*

Remark 9.1.2 *For a real system, e.g., a differential robot, x represents the position and $\ddot{x}$ the acceleration. Assumption 9.1.2 is reasonable in the sense that the acceleration is bounded due to the physical limitation of the agent. For the E-puck robot used in our test (Figure 9.1), the maximum acceleration is set to* 2.5 cm/s^2.

Let $\mathcal{F}(t) = [\mathcal{F}_1(t), \mathcal{F}_2(t), \cdots, \mathcal{F}_N(t)]^T$ describe a time-varying formation structure of the agent network in a reference coordinate frame. The objective of this chapter is to design a fixed-time formation protocol for each agent such that for any initial condition $x_i(0)$, $i = 1, 2, \cdots, N$, there exists a formation settling time $T(x_0) \in [0, \infty)$ satisfying

$$\begin{cases} \lim_{t \to T(x_0)} |(x_i(t) - \mathcal{F}_i(t)) - (x_j(t) - \mathcal{F}_j(t))| \to 0, \\ x_i(t) - x_j(t) = \mathcal{F}_i(t) - \mathcal{F}_j(t), \ \forall t \geq T(x_0). \end{cases} \tag{9.2}$$

Assumption 9.1.3 *The formation vector* $\mathcal{F}(t) = \left[\mathcal{F}_1^T(t), \mathcal{F}_2^T(t), \cdots, \mathcal{F}_N^T(t)\right]^T$ *is bounded with* $\mathcal{F}_i(t)$, $\forall\ i = 1,\ 2,\ \cdots,\ N$ *continuously differentiable, i.e.,* $\|\mathcal{F}(t)\| \leq c_0$, *where* c_0 *is a positive constant.*

Assumption 9.1.4 *The network connection between the robot is undirected.*

Remark 9.1.3 *For an undirected graph* $\mathcal{G}$, *if it is connected, the Laplacian matrix* $\mathcal{L}$ *has a single zero eigenvalue with* $\mathcal{L}1_N = 0$, *and all the other eigenvalues of* $\mathcal{L}$ *are real and positive. Furthermore, we have the following properties for the Laplacian matrix* $\mathcal{L}$ *from [125]:*

1. $x^T\mathcal{L}x = \frac{1}{2}\sum_{i=1}^N \sum_{j=1}^N a_{ij}(x_j - x_i)^2$;

2. $\lambda_2(\mathcal{L}) = \min_{x\neq 0, 1_N^T x = 0} \frac{x^T\mathcal{L}x}{x^T x}$;

3. $x^T\mathcal{L}x \geq \lambda_2(\mathcal{L})x^T x$, *where* $\lambda_2(\mathcal{L})$ *is the smallest positive eigenvalue of* $\mathcal{L}$.

9.1.2 Preliminaries

Lemma 9.1.1 *[9] For a system* $\dot{\chi}(t) = h(\chi, t)$, $\chi(0) = \chi_0$, *in which* $h(\chi) : \mathbb{R}^n \to \mathbb{R}^n$ *is continuous and* $h(0) = 0$. *If* $V(\chi) : \mathbb{R}^n \to \mathbb{R}$ *is a continuously differentiable positive definite function and there exist real numbers* $\varepsilon > 0$, $\lambda \in (0, 1)$ *such that* $\dot{V}(\chi) + \varepsilon V^{\lambda}(\chi) \leq 0$, *then, the origin is a globally finite-time stable equilibrium of the system and the settling time is* $T(\chi_0) \leq \frac{1}{\varepsilon(1-\lambda)} V^{(1-\lambda)}(\chi_0)$.

Lemma 9.1.2 *[203, 205] For a scalar system* $\dot{\chi}(t) = \mathfrak{u}(t)$, $\chi(0) = 0$, *if the controller is chosen as* $\mathfrak{u}(t) = -\alpha\chi^{2-\frac{m}{n}}(t) - \beta\chi^{\frac{m}{n}}(t)$ *with* $\alpha, \beta > 0$ *and* m, n *are both positive odd integers satisfying* $m < n$, *the closed-loop system is fixed-time stable at the origin with globally bounded settling time* $T(\chi_0) \leq \frac{n\pi}{2\sqrt{\alpha\beta(n-m)}}$.

Lemma 9.1.3 *[208] Let* $\beta_1, \beta_2, \cdots, \beta_N \geq 0$ *and* $0 < q < 1$. *Then* $\sum_{j=1}^N \beta_j^q \geq (\sum_{j=1}^N \beta_j)^q$.

9.2 Fixed-Time Formation Control

9.2.1 Fixed-Time Formation with Undirected Topology

To deal with the input delay, one basic idea is to predict the evolution of state variable for the delay period and then use the predicted state for control [3, 161]. For multi-robot systems (9.1), we propose a new state $y(t)$ to predict the state x at time $t+h$

$$y_i(t) = x_i(t) - \mathcal{F}_i(t) + \int_t^{t+h} \Big[u_i(\tau - h) - \dot{\mathcal{F}}_i(\tau)\Big] \mathrm{d}\tau. \tag{9.3}$$

Differentiating $y_i(t)$ against time yields

$$\begin{aligned}\dot{y}_i(t) &= u_i(t-h-\Delta h_i) - \dot{\mathcal{F}}_i(t) + u_i(t) - u_i(t-h) - \dot{\mathcal{F}}_i(t+h) + \dot{\mathcal{F}}_i(t)\\ &= u_i(t) + \tilde{u}_i(t) - \dot{\mathcal{F}}_i(t+h),\end{aligned} \tag{9.4}$$

where $\tilde{u}_i(t) = u_i(t-h-\Delta h_i) - u_i(t-h)$, which represents the uncertain term remains in the transformed system. Before we start the fixed-time formation analysis, a result on the bound of $\tilde{u}_i(t)$ is given first.

Lemma 9.2.1 *If Assumptions 9.1.1 and 9.1.2 hold, the uncertain term $\tilde{u}_i(t)$ in the transformed system (9.4) is bounded and a bound can be established as $0 \le |\tilde{u}_i(t)| \le R_1R_2$.*

Proof 27 *From system (9.1), we have $\tilde{u}_i(t) = \dot{x}_i(t) - \dot{x}_i(t+\Delta h_i) = -\int_t^{t+\Delta h_i} \ddot{x}(\tau)\mathrm{d}\tau$ and $|\tilde{u}_i(t)| = |\int_t^{t+\Delta h_i} \ddot{x}(\tau)\mathrm{d}\tau| \le \int_t^{t+R_1} R_2\mathrm{d}\tau \le R_1R_2$, for all $t \in [0, +\infty)$, which completes the proof.*

The signal collected by agent i about the information of its neighbouring agents is given by

$$\zeta_i(t) = \sum_{j=1}^{N} a_{ij}(y_i(t) - y_j(t)),\ i = 1, 2, \cdots, N.$$

Based on the above preparation, now we present a nonlinear formation protocol for each agent:

$$u_i(t) = -\alpha\zeta_i^{2-\frac{m}{n}}(t) - \beta\zeta_i^{\frac{m}{n}}(t) - \gamma\mathrm{sgn}(\zeta_i(t)) + \dot{\mathcal{F}}_i(t+h), \tag{9.5}$$

where $\alpha, \beta, m, n > 0$ are a set of design parameters, γ is a number satisfying $\gamma \ge R_1R_2$, $\mathrm{sgn}(\cdot)$ denotes the signum function. In particular, we have the following assumption.

Assumption 9.2.1 *m, n are both positive odd integers satisfying $m < n$.*

Theorem 15 *For input-delayed multi-robot systems (9.1) with the associated undirected graph $\mathcal{G}$ being connected, the fixed-time formation problem can be solved by controller (9.5). Furthermore, the settling time satisfies*

$$T(x_0) \le T_{\max} := \frac{\pi n N^{\frac{n-m}{4n}}}{2\sqrt{\alpha\beta}\lambda_2(\mathcal{L})(n-m)} + h.$$

Proof 28 *Putting (9.4) and (9.5) together, we obtain* $\dot{y}_i(t) = -\alpha\left(\zeta_i(t)\right)^{2-\frac{m}{n}} - \beta\left(\zeta_i(t)\right)^{\frac{m}{n}} + \tilde{u}_i(t) - \gamma\mathrm{sgn}\left(\zeta_i(t)\right)$. *Let* $y(t) = \left[y_1(t), y_2(t), \cdots, y_N(t)\right]^T$ *and consider the following semi-positive definite function*

$$V(y(t)) = \frac{1}{2}y^T\mathcal{L}y = \frac{1}{4}\sum_{i=1}^{N}\sum_{j=1}^{N}a_{ij}(y_i(t) - y_j(t))^2. \tag{9.6}$$

Since $\mathcal{G}$ *is connected, zero is a simple eigenvalue of* $\mathcal{L}$, *which implies that* $V(y(t)) = 0$ *if and only if* $y(t) \in \mathrm{span}\{\mathbf{1}_N\}$ *which implies that* $\lim_{t\to\infty}(y_i(t) - y_j(t)) = 0$. *From (9.3) and (9.5), it is concluded that* $\lim\limits_{t\to\infty}(y_i(t) - y_j(t)) = 0 \implies \lim\limits_{t\to\infty}(x_i(t) - x_j(t)) = \mathcal{F}_i(t) - \mathcal{F}_j(t),\ \forall i, j \in N$. *Differentiating* $V(y)$ *against time yields*

$$\begin{aligned}
\dot{V}(y) =& \sum_{i=1}^{N}\frac{\partial V(y)}{\partial y_i}\dot{y}_i(t) \\
=& -\alpha\sum_{i=1}^{N}\zeta_i^{\frac{3n-m}{n}}(t) - \beta\sum_{i=1}^{N}\zeta_i^{\frac{m+n}{n}}(t) - \gamma\sum_{i=1}^{N}|\zeta_i(t)| + \sum_{i=1}^{N}\tilde{u}_i(t)\left(\zeta_i(t)\right) \\
\leq& -\alpha\sum_{i=1}^{N}\left(\zeta_i^2(t)\right)^{\frac{3n-m}{2n}} - \beta\sum_{i=1}^{N}\left(\zeta_i^2(t)\right)^{\frac{m+n}{2n}} - (\gamma - R_1R_2)\sum_{i=1}^{N}|\zeta_i(t)|,
\end{aligned} \tag{9.7}$$

where $\frac{\partial V(y)}{\partial y_i} = \sum_{j=1}^{N}a_{ij}(y_i(t) - y_j(t))$, *due to the symmetry of* $\mathcal{A}$, *is inserted to derive the second equality.*

With Lemmas 9.1.3 and 9.2.1, we have

$$\begin{aligned}
\dot{V}(y) \leq& -\alpha N^{\frac{n-m}{2n}}\left(\sum_{i=1}^{N}\left(\sum_{j=1}^{N}a_{ij}(y_i(t) - y_j(t))\right)^2\right)^{\frac{3n-m}{2n}} \\
&- \beta\left(\sum_{i=1}^{N}\left(\sum_{j=1}^{N}a_{ij}(y_i(t) - y_j(t))\right)^2\right)^{\frac{m+n}{2n}}.
\end{aligned} \tag{9.8}$$

Since $\mathcal{L}$ *is semi-positive definite, we can obtain a matrix* $H \in \mathbb{R}^{N\times N}$ *such that* $\mathcal{L} = H^TH$, *and thus*

$$\begin{aligned}
&\frac{\sum_{i=1}^{N}\left(\sum_{j=1}^{N}a_{ij}(y_i(t) - y_j(t))\right)^2}{V(y)} = \frac{y^T\mathcal{L}^T\mathcal{L}y}{\frac{1}{2}y^T\mathcal{L}y} \\
=& \frac{2y^TH^THH^THy}{y^TH^THy} = \frac{2y^TH^T\mathcal{L}^THy}{y^TH^THy} \geq 2\lambda_2(\mathcal{L}).
\end{aligned} \tag{9.9}$$

Putting (9.8) and (9.9) together, we have

$$\begin{aligned}\dot{V}(y) \leq & -\alpha N^{\frac{n-m}{2n}} (2\lambda_2(\mathcal{L})V)^{\frac{3n-m}{2n}} - \beta(2\lambda_2(\mathcal{L})V)^{\frac{m+n}{2n}} \\ = & -\left(\alpha N^{\frac{n-m}{2n}} (2\lambda_2(\mathcal{L})V)^{\frac{n-m}{n}} + \beta\right)(2\lambda_2(\mathcal{L})V)^{\frac{m+n}{2n}}.\end{aligned}$$

If $V(y) \neq 0$*, we define* $\xi = \sqrt{2\lambda_2(\mathcal{L})V(y)}$*. Then, we obtain* $\dot{\xi} = \lambda_2(\mathcal{L})\dot{V}(y)/\sqrt{2\lambda_2(\mathcal{L})V(y)}$ *and* $\dot{\xi} = -\alpha N^{\frac{n-m}{2n}}\lambda_2(\mathcal{L})\xi^{\frac{2n-m}{n}} - \beta\lambda_2(\mathcal{L})\xi^{\frac{m}{n}}$*. Based on Lemma 9.1.2, we have* $\lim_{t\to T(y_0)} V(y) = 0$ *with the settling time bounded by*

$$\begin{aligned}T(y_0) = & \frac{nN^{\frac{n-m}{4n}}}{\sqrt{\alpha\beta}\lambda_2(\mathcal{L})(n-m)} \tan^{-1}\left(N^{\frac{n-m}{4n}}\sqrt{\frac{\alpha}{\beta}}V(y(0))\right) \\ \leq & \frac{\pi n N^{\frac{n-m}{4n}}}{2\sqrt{\alpha\beta}\lambda_2(\mathcal{L})(n-m)}. \end{aligned} \tag{9.10}$$

The fixed-time consensus of the transformed system (9.4) is achieved.

From (9.10), we know that $y(t) \in \mathrm{span}\{\mathbf{1}_N\}$ *within fixed time* $\frac{\pi n N^{\frac{n-m}{4n}}}{2\sqrt{\alpha\beta}\lambda_2(\mathcal{L})(n-m)}$*. Therefore, we have* $x_i(t) - \mathcal{F}_i(t) = y_i(t) - \int_t^{t+h}[u_i(\tau - h) - \dot{\mathcal{F}}_i(\tau)]\mathrm{d}\tau$ *and* $u_i(t) = \dot{\mathcal{F}}_i(t+h)$ *for all* $t \geq \frac{\pi n N^{\frac{n-m}{4n}}}{2\sqrt{\alpha\beta}\lambda_2(\mathcal{L})(n-m)}$*. It is derived that* $x_i(t) - \mathcal{F}_i(t) = y_i(t)$ *for all* $t \geq \frac{\pi n N^{\frac{n-m}{4n}}}{2\sqrt{\alpha\beta}\lambda_2(\mathcal{L})(n-m)} + h$*. Thus, we have* $T(x_0) \leq \frac{\pi n N^{\frac{n-m}{4n}}}{2\sqrt{\alpha\beta}\lambda_2(\mathcal{L})(n-m)} + h$*, and*

$$\begin{cases} \lim_{t\to T(x)} |(x_i(t) - \mathcal{F}_i(t)) - (x_j(t) - \mathcal{F}_j(t))| \to 0, \\ x_i(t) - x_j(t) = \mathcal{F}_i(t) - \mathcal{F}_j(t), \ \forall t \geq T(x), \end{cases}$$

which completes the proof.

Remark 9.2.1 *The estimation for settling time* $T(x_0)$ *is only related to the delay* h*, the design parameters* m *and* n*, the algebraic connectivity* $\lambda_2(\mathcal{L})$ *of the graph Laplacian and the group order* N*. Therefore, the upper-bound* $T_{\max}$ *of the formation settling time can be designed independent of the initial conditions.*

9.2.2 Fixed-Time Formation with Directed Topology

In the previous subsection, a control protocol for fixed-time formation is developed under undirected topology. To reduce the communication burden and facilitate the practical applications, in this subsection, we will develop a fixed-time formation protocol based on directed local neighbour-to-neighbour interaction.

Assumption 9.2.2 *The directed graph* $\mathcal{G}(A)$ *is strongly connected.*

Definition 13 *[110] Let $\mathcal{G}(A) = \{\mathcal{V}, \mathcal{E}, A\}$ be weighted digraph. Let $\tilde{\mathcal{E}}$ be the set of reverse edges of $\mathcal{G}(A)$ obtained by reversing the order of nodes of all pair in $\mathcal{E}$. The mirror of $\mathcal{G}(A)$ denoted by $\mathcal{G}(\hat{A}) = \{\mathcal{V}, \hat{\mathcal{E}}, \hat{A}\}$ with the same set of nodes as $\mathcal{G}(A)$, the set of edges $\hat{\mathcal{E}} = \mathcal{E} \cup \tilde{\mathcal{E}}$, and the symmetric adjacent matrix $\hat{A} = [\hat{a}_{ij}]_{N\times N}$ with elements $\hat{a}_{ij} = \hat{a}_{ji} = (\delta_i a_{ij} + \delta_j a_{ji})/2$.*

Lemma 9.2.2 *[212] Let $\mathcal{G}(A)$ be weighted digraph with Laplacian matrix $\mathcal{L}$. Then, $\hat{\mathcal{L}} = \left(\Delta\mathcal{L} + \mathcal{L}^T\Delta\right)/2$ is a valid Laplacian matrix for $\mathcal{G}(\hat{A})$, where $\Delta = \mathrm{diag}\{\delta_1, \delta_2, \cdots, \delta_N\}$ is the diagonal matrix whose diagonal entries are given by δ_i, $i = 1, 2, \cdots, N$.*

The signal collected by agent i about the information of its neighbouring agents is given by

$$\hat{\zeta}_i(t) = \sum_{j=1}^{N} \hat{a}_{ij}(y_i(t) - y_j(t)), \ i = 1, 2, \cdots, N.$$

A nonlinear protocol that solves the fixed-time formation problem for the multi-robot systems with directed topology can be proposed as

$$u_i(t) = -\alpha\hat{\zeta}_i^{2-\frac{m}{n}}(t) - \beta\hat{\zeta}_i^{\frac{m}{n}}(t) - \gamma\mathrm{sgn}(\hat{\zeta}_i(t)) + \dot{\mathcal{F}}_i(t+h), \tag{9.11}$$

where $\alpha, \beta, \gamma, m, n$ are defined in (9.5), y_i and y_j are given in (9.3), and $\hat{a}_{ij}$ is given in Definition 13.

Theorem 16 *With Assumption 9.2.2, the fixed-time formation problem of multi-robot systems (9.1) can be solved by the algorithm (9.11). Furthermore, the settling time satisfies $T(x) \leq \frac{\pi n N^{\frac{n-m}{4n}}}{2\sqrt{\alpha\beta}\lambda_2(\hat{\mathcal{L}})(n-m)} + h$, where $\lambda_2(\hat{\mathcal{L}})$ is the smallest positive eigenvalue of $\hat{\mathcal{L}}$.*

Proof 29 *The proof is similar to that of Theorem 1 and hence omitted.*

Remark 9.2.2 *From (9.2) and Assumption 9.1.3, it can be seen that the consensus of all robots can be achieved if $\mathcal{F}(t) \equiv 0$. Thus, the fixed-time consensus problem studied in [208] can be viewed as a special case of the fixed-time formation problem. If $\Delta h_i \equiv 0$, $\forall\ i = 1, 2, \cdots, N$, the delay problem under consideration is equivalent to the identical delay case considered in [32] and [161].*

9.3 A Numerical Example

This section presents some simulation results on convergence of six agents with $\mathcal{F}(t) = 0$. The connection is shown in Figure 9.2. In the simulation, nonzero $a_{ij} = 1$, $\alpha = \beta = 4$, $m = 7$ and $n = 9$ for each protocol, the

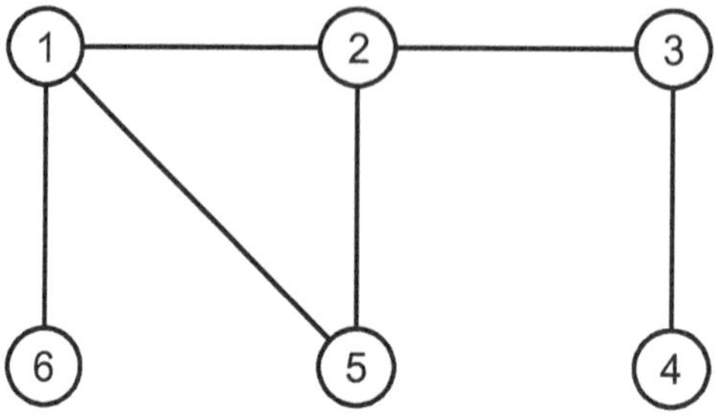

FIGURE 9.2: Communication topology.

time delay of the system $h = 0.3, \Delta h_i = 0.05\sin(i \cdot t), \forall\ i = 1, 2, \cdots, 6$. The adjacency matrix A is given by

$$A = \begin{bmatrix} 0 & 1 & 0 & 0 & 1 & 1 \\ 1 & 0 & 1 & 0 & 1 & 0 \\ 0 & 1 & 0 & 1 & 0 & 0 \\ 0 & 0 & 1 & 0 & 0 & 0 \\ 1 & 1 & 0 & 0 & 0 & 0 \\ 1 & 0 & 0 & 0 & 0 & 0 \end{bmatrix}.$$

The Laplacian matrix is given by

$$\mathcal{L} = \begin{bmatrix} 3 & -1 & 0 & 0 & -1 & -1 \\ -1 & 3 & -1 & 0 & -1 & 0 \\ 0 & -1 & 2 & -1 & 0 & 0 \\ 0 & 0 & -1 & 1 & 0 & 0 \\ -1 & -1 & 0 & 0 & 2 & 0 \\ -1 & 0 & 0 & 0 & 0 & 1 \end{bmatrix}.$$

The eigenvalues of $\mathcal{L}$ are $\{0,\ 0.4131,\ 1.1369,\ 2.3595,\ 3.6977,\ 4.3928\}$, and zero is a simple eigenvalue. We obtain that $\lambda_2(\mathcal{L}) = 0.4131$. The estimated upper bound of the settling time for protocol (9.5) is 5.0231. Consider two initial conditions: (1) $x(0) = [10, -20, -3, 9, 4, -30]^T$ and (2) $x(0) = [100, -200, -30, 90, 40, -300]^T$.

To demonstrate the communication and input delay effect in the consensus control, first we will show the simulation results with the controllers designed in [208]. Figures 9.3 and 9.4 show that the consensus cannot be achieved by the controllers given in [208] even if the delays are identical and very small with $h = 0.03s$.

From the results shown in Figures 9.5 and 9.6, it is demonstrated that the convergence performance is guaranteed with the controller designed in this chapter under different initial conditions. In addition, the numerical results given in Figure 9.5 show that the settling time of the consensus protocol is about 5 s, which demonstrates the correctness of the estimation derived in Theorem 15.

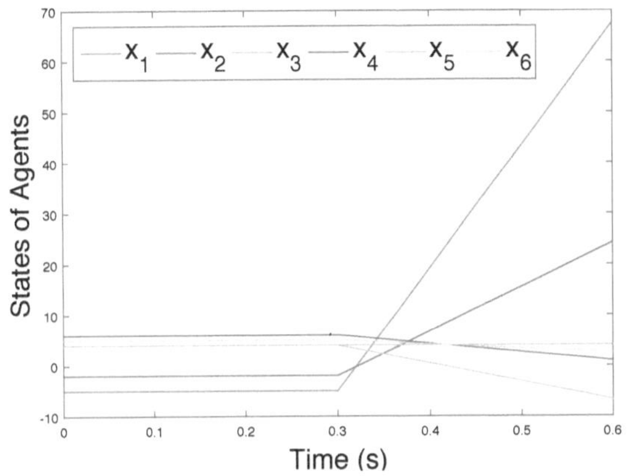

FIGURE 9.3: State trajectories of 6 agents under the controller designed in [208] with $\Delta h_i = 0, h = 0.03$ (Case 1).

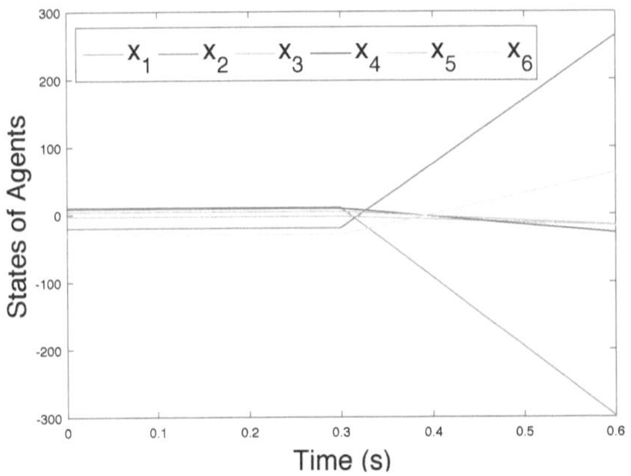

FIGURE 9.4: State trajectories of 6 agents under the controller designed in [208] with $\Delta h_i = 0, h = 0.03$ (Case 2).

9.4 Experiment Validation

9.4.1 Experimental Platform

An experimental testbed, see Figure 9.7, is designed to demonstrate the performance of formation protocols with real robotic systems. In this testbed, a Linux-based host computer (2.7-GHz clock speed, dual processor, 4-GB RAM, and equipped with *Robot Operating Systems*) receives position informa-

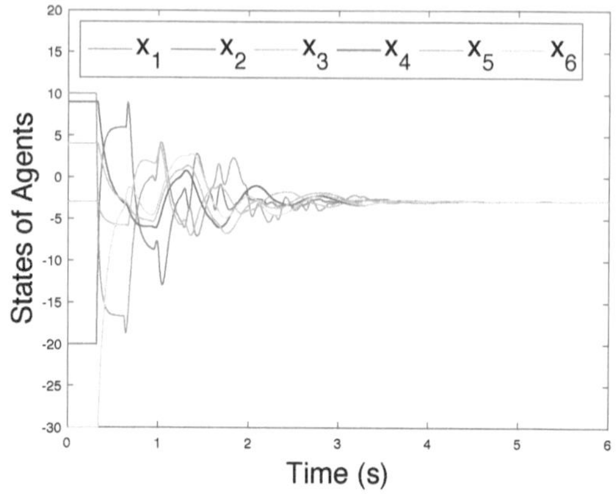

FIGURE 9.5: State trajectories of 6 agents with $h = 0.3, \Delta h_i = \sin(i*t), i = 1, 2, \cdots, 6$ (Case 1).

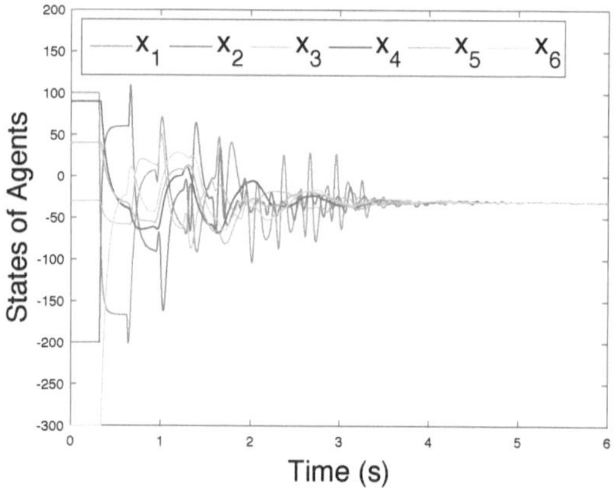

FIGURE 9.6: State trajectories of 6 agents with $h = 0.3, \Delta h_i = \sin(i*t), i = 1, 2, \cdots, 6$ (Case 2).

tion of four E-puck mobile robots from a camera (Logitech C525, HD 720p). The camera used here is to localize the robots within the workspace area (100 cm×100 cm). Some static landmarks are used by the camera to calibrate the workspace coordinates. Images captured by the camera are then processed by utilizing image processing software Swarmcon to provide the positions of robots [63]. The way this software locates the robots is by tracking the unique patterns attached to them. The positions obtained are subsequently applied to the proposed control law to produce the control inputs. Finally, these inputs are transmitted over the network via Bluetooth connection.

FIGURE 9.7: Experimental Platform

In this system, an inherent time delay whose value is around $0.15s$ to $0.195s$ exists in the input due to camera latency, image processing, decision-making process and Bluetooth communication [121]. Since the Bluetooth broadcast to the robots is done sequentially, the delays are in general not identical when communicating with different robots. In this case, we set $h = 0.5s$ and use the inherent delay as the uncertainty term $\Delta h_i \in [0.15s, 0.195s]$, $\forall i = 1, 2, 3, 4$.

9.4.2 Linearization-Based Kinematic Model of E-Puck Robot

Let $(x,\ y)$, θ, v and ω denote the Cartesian positions, the orientation, the linear and the angular velocity of the robot. The kinematic model of the

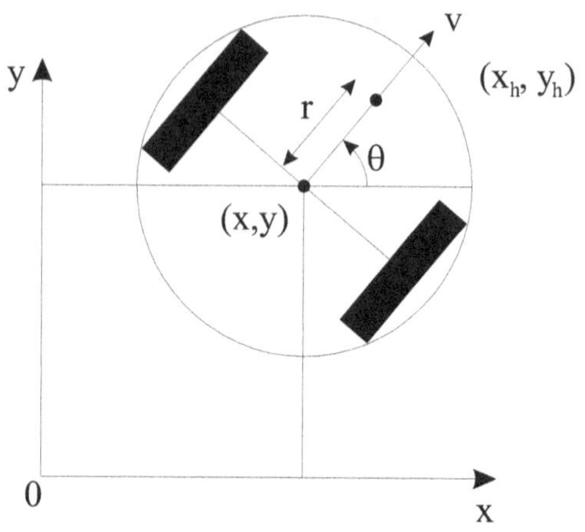

FIGURE 9.8: Coordinates of the differential drive robot.

E-puck robot is described as

$$\begin{bmatrix} \dot{x} \\ \dot{y} \\ \dot{\theta} \end{bmatrix} = \begin{bmatrix} \cos\theta & 0 \\ \sin\theta & 0 \\ 0 & 1 \end{bmatrix} \begin{bmatrix} v \\ \omega \end{bmatrix}. \tag{9.12}$$

Figure 9.8 describes the position of the robot in the global inertial frame. The robot is located at P in the frame which is represented by $p = [x,\ y]^T$. Consider the nonholonomic kinematic constraint, the head position $\hat{p} = [x_h,\ y_h]^T$ as shown in Figure 9.8 is used to replace the inertial position. It can be verified that

$$\begin{bmatrix} \dot{x}_h \\ \dot{y}_h \\ \dot{\theta} \end{bmatrix} = \begin{bmatrix} \cos\theta & -r\sin\theta \\ \sin\theta & r\cos\theta \\ 0 & 1 \end{bmatrix} \begin{bmatrix} v \\ \omega \end{bmatrix}, \tag{9.13}$$

where r is the distance between the head position and the inertial position. By letting $u = [u_x,\ u_y]^T$ and $\eta = [v,\ \omega]^T$, the linearization-based kinematic model can be developed as $\eta(t) = Hu(t)$ and $u(t) = \dot{\hat{p}}(t)$ with $H = \begin{bmatrix} \cos\theta & \sin\theta \\ -\frac{1}{r}\sin\theta & \frac{1}{r}\cos\theta \end{bmatrix}$. Note that in the following experiments, we choose $r = 2.5$ cm.

9.4.3 Static Formation with Four E-Puck Robots

The aim of this experiment is to show that the proposed control law will drive the robots to form a formation whose pattern is described as

$$\mathcal{F}_i = \begin{bmatrix} R\cos(\frac{2\pi(i-1)}{N} + \psi) \\ R\sin(\frac{2\pi(i-1)}{N} + \psi) \end{bmatrix}, \quad i = 1, 2, 3, 4,$$

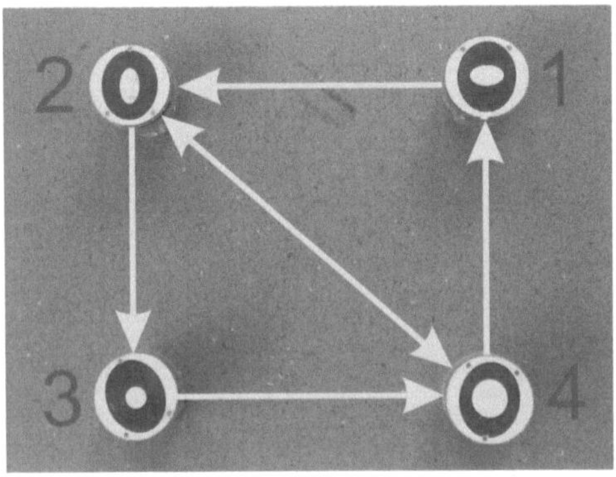

FIGURE 9.9: Directed communication topology.

where R is the distance between the centre of the formation frame and the robot, ψ is the desired angle between x-axis and the first robot with respect to the centre of the formation. In this test, we choose $R = 30\,\text{cm}$ and $\psi = (\pi/4)\,\text{rad}$.

Suppose the directed graph in Figure 9.9 is used to model the information exchange among robots. Accordingly, the Laplacian matrix $\mathcal{L}$ is given by

$$\mathcal{L} = \begin{bmatrix} 1 & 0 & 0 & -1 \\ -1 & 2 & 0 & -1 \\ 0 & -1 & 1 & 0 \\ 0 & -1 & -1 & 2 \end{bmatrix},$$

and

$$\hat{\mathcal{L}} = \begin{bmatrix} 1 & -0.5 & 0 & -0.5 \\ -0.5 & 2 & -0.5 & -1 \\ 0 & -0.5 & 1 & -0.5 \\ -0.5 & -1 & -0.5 & 2 \end{bmatrix}.$$

The eigenvalues of the Laplacian matrix $\hat{\mathcal{L}}$ are $\{0,\ 1.0,\ 2.0,\ 3.0\}$, and zero is a simple eigenvalue. By setting $\alpha = \beta = 0.5$, $m = 19$ and $n = 21$ for each protocol, the estimated upper bound of the settling time for protocol (9.11) is 34.5951 s.

The experimental results of this scenario are displayed in Figures 9.10, 9.11 and 9.12. Fig. 9.10 shows the formation achieving process at $t = 0\,\text{s}, 6\,\text{s}, 12\,\text{s}, 18\,\text{s}, 24\,\text{s}, 30\,\text{s}$. In more detail, Figure 9.11 visualizes the trajectories of robots from the initial poisitons represented by dots. The triangles representing the position of robots at time $t = 10\,\text{s}$ show that the robots are in the process of forming the rectangle formation, while the squares represent the final position of robots. Furthermore, the final positions are reached before $t = 20\,\text{s}$ as seen in Figure 9.12. It shows the robustness of the estimated formation settling time.

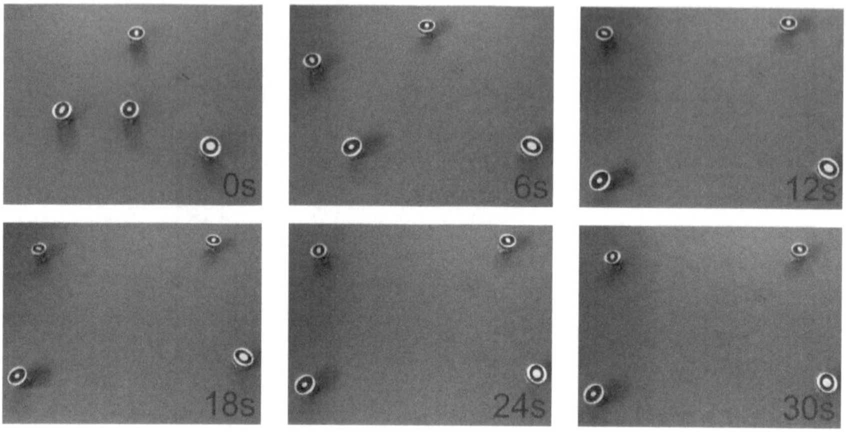

FIGURE 9.10: Static formation with four robots.

Remark 9.4.1 *Notice that saturations of linear and angular velocities have to be enforced in the experiment due to the physical limitations of the robots. Compared with the simulation setting in Section 9.3, bigger value of m/n is chosen in the test to avoid possible saturation. Furthermore, the approximation method introduced in [197] is used to deal with the implementation of the integral term in (9.3).*

9.4.4 Time-Varying Formation with Four E-Puck Robots

For time-varying formations, four robots are commanded to move on the $x-y$ plane to form a time-varying formation in fixed-time. The formation is defined by

$$\mathcal{F}_i(t) = \begin{bmatrix} R\cos(\frac{2\pi t}{T} + \frac{2\pi(i-1)}{N} + \psi) \\ R\sin(\frac{2\pi t}{T} + \frac{2\pi(i-1)}{N} + \psi) \end{bmatrix}, \quad i = 1, 2, 3, 4,$$

which represents the periodic time-varying formation and rotating around the formation centre with period T. In this case, we choose $R = 30\,\text{cm}$ and $\psi = 0\,\text{rad}$. By doing a simple mathematical analysis, if the maximum velocity of robot is $5\,\text{cm/s}$, the value of T must be larger or equal to $25.1327\,\text{s}$ in order to make the robots follow the desired path. In this case, T is chosen to be $50.2655\,\text{s}$. The directed graph in Figure 9.9 is used to model the information exchange among robots. By setting $\alpha = \beta = 1$, $m = 19$ and $n = 21$ for each protocol, the estimated upper bound of the settling time for protocol (9.11) is $34.5951\,\text{s}$.

Figures. 9.13, 9.14 and 9.15 depict the experimental results of the proposed time-varying formation. In Figure 9.13, we can see that the robots are initially deployed in random positions within the workspace. After 20 s, the robots rotate around the centre of formation. Figure 9.14 shows the trajectories of

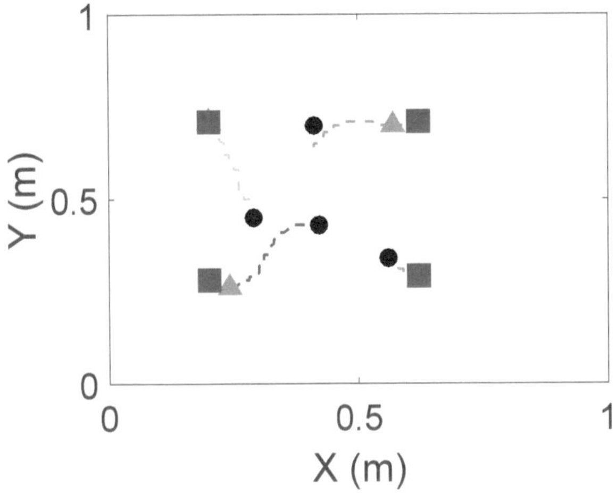

FIGURE 9.11: Positions of each robots while forming a rectangle formation.

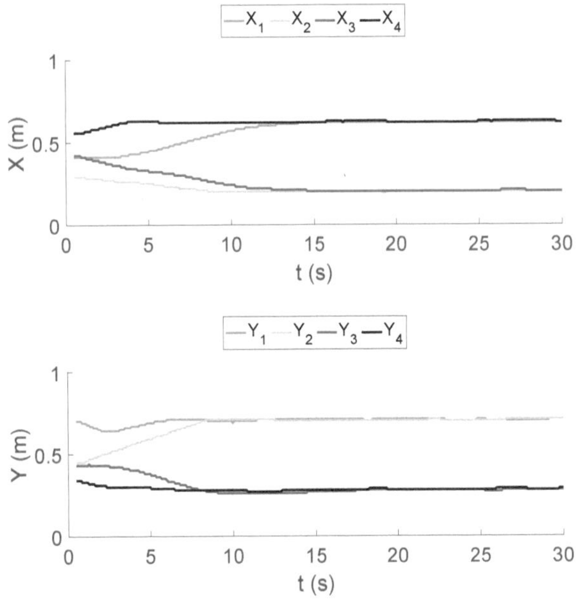

FIGURE 9.12: Trajectories of the states with respect to time.

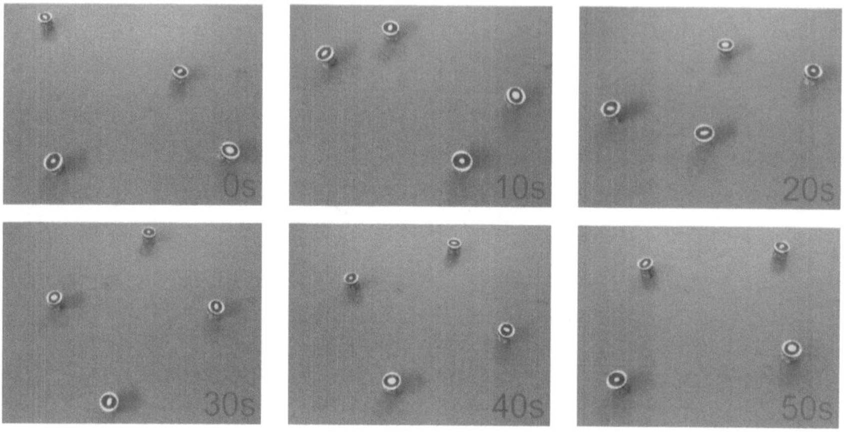

FIGURE 9.13: Time-varying formation with four robots.

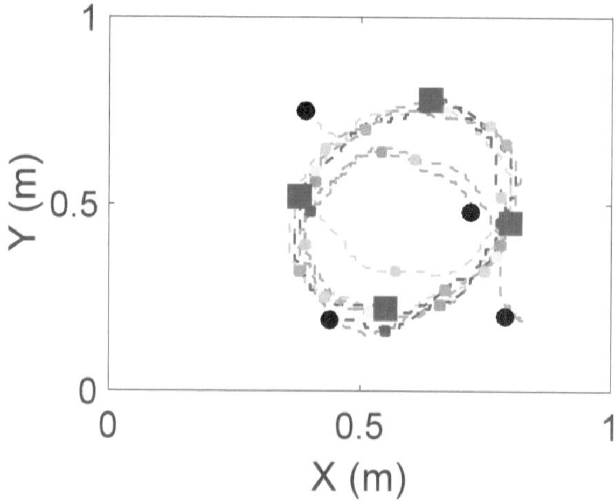

FIGURE 9.14: Robots rotating around the centre of formation.

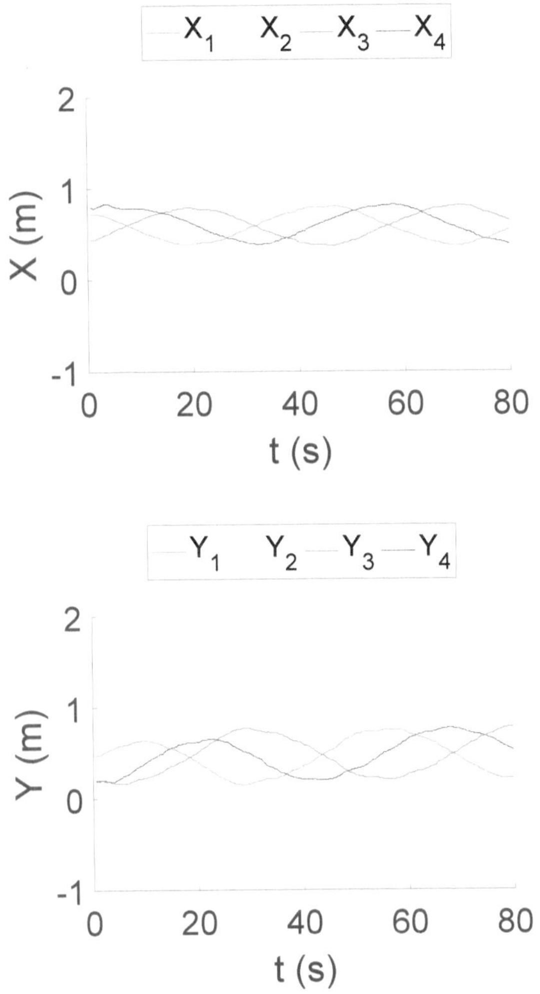

FIGURE 9.15: Trajectories of the states with respect to time.

robots starting from the initial positions as pointed by the dots. The robots keep rotating around the centre of formation during the experiment, and then stop at some positions pointed by squares at the end time. In Figure 9.15, the trajectories of robots with respect to time start to form sinusoidal waves before t = 30 s.

9.5 Conclusions

This chapter has investigated the impact of non-identical input delay in fixed-time formation control for multi-robot systems. This input delay may represent the communication delay and the processing and connecting time for the packets arriving at each agent. Based on the predictor strategy, a state transformation is applied for each agent and the extra terms remain in the transformed systems are carefully considered. A new class of fixed-time nonlinear formation protocols are constructed and a complete stability analysis is presented in a systematic framework of Lyapunov functionals. The upper bound of the formation settling time is independent of the system initial conditions such that the convergence time can be designed or estimated off-line. Finally, both simulation and experimental results are presented to show the effectiveness of the protocols proposed in this chapter.

9.6 Notes

This chapter extends an idea in [208, 212] and develops fixed-time formation protocols for multi-robot systems subject to delay constraints. The main materials of this chapter are based on [157, 165]. It is worth mentioning that the fixed-time formation protocols proposed in this chapter are designed for first-order integrator multi-agent systems. The results in this chapter can be extended to second/high-order multi-agent systems by following the procedures shown in [203] and [204].

Chapter 10

Cascade Structure Predictive Observer Design for Consensus Control with Applications to UAVs Formation Flying

This chapter investigates the long input delay problem in consensus control of multi-agent systems with directed graph. First, cascade structure predictive observers are designed to obtain the future state for each agent. Then, adaptive technique based consensus controllers are designed for the agents to make sure that any global information is not needed. Lyapunov-Krasovskii functionals are used for the delay upper bound analysis and sufficient conditions are derived to guarantee that the consensus errors converge to zero asymptotically. Finally, the proposed controllers are verifed through numerical simulation and UAVs formation flying tests.

10.1 Problem Formulation

Consider the following multi-agent systems in a leader-follower struture

$$\dot{x}_i(t) = Ax_i(t) + Bu_i(t-h), \tag{10.1}$$

where for agent i, $i = 0, 1, \ldots, N$, $x_i \in \mathbb{R}^n$ is the state, $u_i \in \mathbb{R}^q$ is the control input, $A \in \mathbb{R}^{n\times n}$, and $B \in \mathbb{R}^{n\times q}$ are constant matrices with (A, B) being controllable, $h > 0$ is a long input delay and uniform across all the agents. Hereafter, the argument t is omitted in some equations to save space.

The aim of this chapter is to design predictive observers and finite-dimensional controllers to guarantee that the state of each follower converges to the leader state for any initial condition of $x_i(0)$, $i = 1, 2, \cdots, N$.

Assumption 10.1.1 *The communication topology $\mathcal{G}$ contains a directed spanning tree.*

Assumption 10.1.2 *The leader is indexed by 0 and has no control input.*

For the nonsingular M-matrix $\mathcal{L}_1$, there exists a positive diagonal matrix G such that

$$G\mathcal{L}_1 + \mathcal{L}_1^T G \geq \rho_0 I, \tag{10.2}$$

for some positive constant ρ_0. G can be constructed by letting $G = \text{diag}\{g_1, g_2, \cdots, g_N\}$, where $g = [g_1, g_2, \cdots, g_N]^T = \left(\mathcal{L}_1^T\right)^{-1} [1, 1, \cdots, 1]^T$.

Next, we introduce the predictor feedback idea. Consider a linear system

$$\dot{x}(t) = Ax(t) + Bu(t-h). \tag{10.3}$$

With predictor feedback [3, 64, 66], the controller can be designed as

$$u(t) = Kx(t+h) = K\left(e^{Ah}x(t) + \int_{t-h}^{t} e^{A(t-\tau)}Bu(\tau)d\tau\right). \tag{10.4}$$

Then, the closed-loop system will be delay-free and can be expressed as

$$\dot{x}(t) = (A + BK)x(t). \tag{10.5}$$

If $A + BK$ is Hurwitz, then the stability problem is solved.

Remark 10.1.1 *However, the controller (10.4) is hard to implement. First, more advanced processors and actuators are needed to take the integral of the past control inputs which increases the cost and difficulty of the implementation. Second, the controller (10.4) is infinite-dimensional and contains distributed delays. As mentioned in [197], the distributed delays can be replaced by the sum of a series of discrete delays with the approximation method. Unfortunately, the approximation method cannot guarantee system stability, even when quite accurate approximation integral laws are used [4, 37].*

To solve this problem, we can design a predictive observer $z(t)$ to estimate $x(t+h)$

$$\dot{z}(t) = Az(t) + Bu(t) + BK\left(z(t-h) - x(t)\right). \tag{10.6}$$

The observer error is defined by $e(t) = x(t) - z(t-h)$. It follows that

$$\dot{e}(t) = Ae(t) + BKe(t-h). \tag{10.7}$$

If $A + BK$ is Hurwitz, for system (10.7), there certainly exists a sufficiently small h such that $\lim_{t\to\infty} e = 0$ (see reference [104] for detail) and the observer state $z(t)$ predicts the state x at time $t+h$ with $z(t) \longrightarrow x(t+h)$. Thus, finite-dimensional controller could be designed as

$$u(t) = Kz(t). \tag{10.8}$$

Then, the stability problem of system (10.3) is solved.

Remark 10.1.2 *This predictive observer idea is first proposed in [8] for single linear systems with input delay and is then extended to single linear systems with both input and state delays in [201]. It should be pointed out that the single predictive observer (10.6) can only deal with small input delay. For a system with long input delay, the stability of the closed-loop system (10.7) cannot be guaranteed.*

10.2 Delay Upper Bound Analysis

Define $z_i(t)$ as the estimation of the state x_i at time $t+h$. For the leader, we have $z_0(t) = e^{Ah}x_0(t)$. For the followers, we construct the state predictive observer as

$$\dot{z}_i(t) = Az_i(t) + Bu_i(t) + BK\left(z_i(t-h) - x_i(t)\right), \tag{10.9}$$

where $i = 1, 2, \cdots, N$. The estimation error is defined by $e_i(t) = x_i(t) - z_i(t-h)$. It follows that

$$\dot{e}_i(t) = Ae_i(t) + BKe_i(t-h), \quad i = 1, 2, \cdots, N. \tag{10.10}$$

Let $e = [e_1^T, e_2^T, \cdots, e_N^T]^T$. Then, we have

$$\dot{e}(t) = \left(I_N \otimes A\right)e(t) + \left(I_N \otimes BK\right)e(t-h). \tag{10.11}$$

Lemma 10.2.1 *For system (10.11), given a scalar h^*, if there exist positive definite matrix W and constants $\beta_1, \beta_2 > 0$ such that*

$$\begin{bmatrix} (WA^T + AW - 2BB^T)/h^* & WA^T & BB^T \\ AW & -\beta_1 W & 0 \\ BB^T & 0 & -\tilde{\beta}^{-1}W \end{bmatrix} < 0, \tag{10.12}$$

where $\tilde{\beta} = \beta_1 + \beta_2 + \beta_2^{-1}$, then system (10.11) is asymptotically stable with $K = -B^TP$, $P = W^{-1}$, for any delay h satisfying $0 \leq h \leq h^*$.

Proof 30 *To show the stability of systems (10.11), a Lyapunov functional candidate is defined as*

$$\begin{aligned} V_0(t) = {} & e^T(t)\left(I_N \otimes P\right)e(t) + \int_{t-h}^{t}\int_{\tau}^{t} r_1\left[e^T(s)\left(I_N \otimes A^TPA\right)e(s)\right]\mathrm{d}s\mathrm{d}\tau \\ & + \int_{t-h}^{t}\int_{\tau-h}^{t} r_2\left[e^T(s)\left(I_N \otimes Q\right)e(s)\right]\mathrm{d}s\mathrm{d}\tau \end{aligned},$$

where $Q = PBB^TPBB^TP$, and r_1, r_2 are scalars. The time derivative of V_0

along the trajectories of (10.11) is

$$\begin{aligned}
\dot{V}_0(t) &= e^T(t)\left[I_N \otimes (PA + A^T P - 2PBB^T P\right] e(t) \\
&\quad -2e^T(t)\left(I_N \otimes PBK\right)\int_{-h}^{0}\left(I_N \otimes A\right) e(t+\tau)\mathrm{d}\tau \\
&\quad -2e^T(t)\left(I_N \otimes PBK\right)\int_{-h}^{0}\left(I_N \otimes BK\right) e(t+\tau-h)\mathrm{d}\tau \\
&\quad -\int_{-h}^{0} r_1\left[e^T(t+\tau)\left(I_N \otimes A^T PA\right) e(t+\tau)\right]\mathrm{d}\tau \\
&\quad -\int_{-h}^{0} r_2\left[e^T(t+\tau-h)\left(I_N \otimes Q\right) e(t+\tau-h)\right]\mathrm{d}\tau \\
&\quad +hr_1 e^T(t)\left(I \otimes A^T PA\right) e(t) + hr_2 e^T(t)\left(I \otimes Q\right) e(t). \qquad (10.13)
\end{aligned}$$

With $K = -B^T P$, we can obtain

$$\begin{aligned}
&-2e^T(t)\left(I_N \otimes PBK\right)\int_{-h}^{0}\left(I_N \otimes A\right) e(t+\tau)\mathrm{d}\tau \\
&-2e^T(t)\left(I_N \otimes PBK\right)\int_{-h}^{0}\left(I_N \otimes BK\right) e(t+\tau-h)\mathrm{d}\tau \\
\leq\; & \int_{-h}^{0} r_1\left[e^T(t+\tau)\left(I_N \otimes A^T PA\right) e(t+\tau)\right]\mathrm{d}\tau \\
&+\int_{-h}^{0} r_2\left[e^T(t+\tau-h)\left(I_N \otimes Q\right) e(t+\tau-h)\right]\mathrm{d}\tau \\
&+\left(hr_1^{-1} + hr_2^{-1}\right) e^T(t)\left(I_N \otimes Q\right) e(t). \qquad (10.14)
\end{aligned}$$

Then, putting (10.13) and (10.14) together, it is derived that

$$\begin{aligned}
\dot{V}_0(t) \;\leq\; & e^T(t)\left[I_N \otimes (PA + A^T P - 2PBB^T P + hr_1 A^T PA\right. \\
&\left. + \left(hr_1^{-1} + hr_2^{-1} + hr_2\right) PBB^T PBB^T P)\right] e(t). \qquad (10.15)
\end{aligned}$$

If inequality (10.12) in Lemma 10.2.1 is satisfied, with Schur complement lemma, we will have the following inequality

$$\begin{aligned}
PA &+ A^T P - 2PBB^T P + hr_1 A^T PA \\
&+ \left(hr_1^{-1} + hr_2^{-1} + hr_2\right) PBB^T PBB^T P < 0. \qquad (10.16)
\end{aligned}$$

where $r_1 = \beta_1^{-1}, r_2 = \beta_2^{-1}$, and $P = W^{-1}$ have been used for this derivation. From (10.15) and (10.16), we will have $\dot{V}_0(t) < 0$. This completes the proof.

Remark 10.2.1 *The conditions shown in (10.12) can be checked by standard LMI routines for a set of fixed values. The algorithm for finding the feasible solution of (10.12) is summarized as follows.*

(1) Fix the value of β_1, β_2 *to some constants* $\tilde{\beta}_1 > 0, \tilde{\beta}_2 > 0$*; make an initial guess for the values of* $\tilde{\beta}_1, \tilde{\beta}_2$.

(2) Solve the LMI equation (10.12) for W *with the fixed values; if a feasible value of* W *cannot be founded, return to step (1) and reset the values of* $\tilde{\beta}_1, \tilde{\beta}_2$.

10.3 Observer and Controller Design

10.3.1 Cascade Predictive Observer Design

If h is much bigger than the delay upper bound h^* found with (10.12), i.e., $h \gg h^*$, the stability of (10.10) cannot be guaranteed and the observer (10.9) for each agent will fail. In such situation, cascade predictive observer design can be used to solve this problem. Let $h_1 = h/m$, where h_1 is a small delay value such that $0 \leq h_1 \leq h^*$ and m is a positive integer. Then, the observer for ith follower can be constructed as

$$\begin{cases} \dot{\tilde{z}}_{ip}(t) = A\tilde{z}_{ip}(t) + Bu_i(t-(p-1)h_1) \\ \qquad\qquad + BK\left(\tilde{z}_{ip}(t-h_1) - \tilde{z}_{i(p+1)}(t)\right), \\ \dot{\tilde{z}}_{im}(t) = A\tilde{z}_{im}(t) + Bu_i(t-(m-1)h_1) \\ \qquad\qquad + BK\left(\tilde{z}_{im}(t-h_1) - x_i(t)\right), \end{cases} \tag{10.17}$$

where $p \in \mathbf{I}[1, m-1]$. Define the prediction error vectors as

$$\begin{cases} \tilde{e}_{ip}(t) = \tilde{z}_{ip}\left(t-(m-p+1)h_1\right) - \tilde{z}_{i(p+1)}\left(t-(m-p)h_1\right), \quad p \in \mathbf{I}[1, m-1], \\ \tilde{e}_{im}(t) = \tilde{z}_{im}\left(t-h_1\right) - x_i(t). \end{cases}$$

Then, we have that

$$\tilde{z}_{i1}\left(t-h\right) = x_i(t) + \tilde{e}_{i1}(t) + \tilde{e}_{i2}(t) + \cdots + \tilde{e}_{im}(t), \tag{10.18}$$

and

$$\begin{cases} \dot{\tilde{e}}_{ip}(t) = A\tilde{e}_{ip}(t) + BK\tilde{e}_{ip}(t-h_1) \\ \qquad\qquad - BK\tilde{e}_{i(p+1)}(t-h_1), \quad p \in \mathbf{I}[1, m-1], \\ \dot{\tilde{e}}_{im}(t) = A\tilde{e}_{im}(t) + BK\tilde{e}_{im}(t-h_1). \end{cases} \tag{10.19}$$

(10.19) is asymptotically stable if the following system

$$\dot{\tilde{e}}_{ip}(t) = A\tilde{e}_{ip}(t) + BK\tilde{e}_{ip}(t-h_1), \quad p \in \mathbf{I}[1, m-1], \tag{10.20}$$

is stable. Therefore, the observer state $\tilde{z}_{i1}$ can obtain the future state of x_i at time $t+h$ with $\tilde{z}_{i1}(t) \longrightarrow x_i(t+h)$.

10.3.2 Distributed Robust Adaptive Controller Design

Based on (10.17), distributed robust adaptive controllers are designed as

$$
\begin{aligned}
u_i(t) &= (\phi_i + \gamma_i)K\eta_i - K\tilde{e}_{i1}(t+(m-1)h_1), && (10.21)\\
\dot{\phi}_i(t) &= \eta_i^T(t)\Gamma\eta_i(t), && (10.22)\\
\eta_i(t) &= \sum_{j=0}^{N} a_{ij}\left(\tilde{z}_{i1}(t) - \tilde{z}_{j1}(t)\right), && (10.23)
\end{aligned}
$$

where $i = 1, 2, \cdots, N$, $\gamma_i(t) = \eta_i^T(t)P\eta_i(t)$, Γ is a parameter to be determined later.

Let $z = [z_{11}^T, z_{21}^T, \cdots, z_{N1}^T]^T$, $\eta = [\eta_1^T, \eta_2^T, \cdots, \eta_N^T]^T$, $\phi = \text{diag}(\phi_1, \phi_2, \cdots, \phi_N)$ and $\gamma = \text{diag}(\gamma_1, \gamma_2, \cdots, \gamma_N)$. Substituting (10.21) into the first equation of (10.17), we have

$$
\begin{aligned}
\dot{\tilde{z}}_{i1} &= A\tilde{z}_{i1} + Bu_i + BK\left(\tilde{z}_{i1}(t-h_1) - \tilde{z}_{i2}\right)\\
&= A\tilde{z}_{i1} + BK\left(\tilde{z}_{i1}(t-h_1) - \tilde{z}_{i2}\right)\\
&\quad + B[(\phi_i + \gamma_i(t))K\eta_i - K\tilde{e}_{i1}(t+(m-1)h_1)]\\
&= A\tilde{z}_{i1} + BK\left(\tilde{z}_{i1}(t-h_1) - \tilde{z}_{i2}\right)\\
&\quad + B(\phi_i + \gamma_i)K\eta_i - BK(\tilde{z}_{i1}(t-h_1) - \tilde{z}_{i2})\\
&= A\tilde{z}_{i1} + B(\phi_i + \gamma_i(t))K\eta_i,
\end{aligned}
$$

and

$$\dot{z}(t) = (I_N \otimes A)\, z(t) + [(\phi(t) + \gamma(t)) \otimes BK]\, \eta(t).$$

Based on (10.23) and Assumption 10.1.1, we obtain that

$$\eta(t) = (\mathcal{L}_1 \otimes I_N)\left(z(t) - \mathbf{1} \otimes z_0(t)\right), \tag{10.24}$$

and

$$\dot{\eta}(t) = [I_N \otimes A + \mathcal{L}_1(\phi(t) + \gamma(t)) \otimes BK]\, \eta(t). \tag{10.25}$$

Theorem 17 *For systems (10.1), the consensus problem can be solved by the observer (10.17) and adaptive protocols (10.21)–(10.23) with $K = -B^TP$, $\Gamma = PBB^TP$ and $\gamma_i(t) = \eta_i^T(t)P\eta_i(t)$, where $P > 0$ is defined in Lemma 10.2.1.*

Proof 31 *We define*

$$V_1(t) = \sum_{i=1}^{N} \frac{1}{2} g_i(2\phi_i + \gamma_i)\gamma_i + \sum_{i=1}^{N} \frac{1}{2} g_i(\phi_i(t) - r_3)^2,$$

where r_3 is a positive scalar. The derivation of V_1 is given as

$$\begin{aligned}\dot{V}_1(t) &= \sum_{i=1}^{N} g_i(\phi_i+\gamma_i)\eta_i^T P\dot{\eta}_i + \sum_{i=1}^{N} g_i(\phi_i+\gamma_i)\dot{\eta}_i^T P\eta_i \\ &\quad + \sum_{i=1}^{N} g_i(\gamma_i+\phi_i-r_3)\dot{\phi}_i. \end{aligned} \tag{10.26}$$

Substituting (10.25) into (10.26), we have

$$\begin{aligned}\dot{V}_1(t) &= \eta^T\left[(\phi+\gamma-r_3I)\,G\otimes\Gamma\right]\eta + \eta^T\left[(\phi+\gamma)G\otimes\left(PA+A^TP\right)\right. \\ &\quad \left.-(\phi+\gamma)\left(G\mathcal{L}_1+\mathcal{L}_1^TG\right)(\phi+\gamma)\otimes\Gamma\right]\eta.\end{aligned}$$

With the leader-follower topology and $G\mathcal{L}_1+\mathcal{L}_1^TG \geq \rho_0 I$ in 10.2, it can be further derived that

$$\begin{aligned}\dot{V}_1(t) &\leq \eta^T\left[(\phi+\gamma)G\otimes\left(PA+A^TP\right)-\rho_0(\phi+\gamma)^2\otimes\Gamma\right]\eta \\ &\quad +\eta^T\left[(\phi+\gamma-r_3I)\,G\otimes\Gamma\right]\eta \\ &\leq \eta^T\left[(\phi+\gamma)G\otimes\left(PA+A^TP+\Gamma\right)-\left[\rho_0(\phi+\gamma)^2+r_3G\right]\otimes\Gamma\right]\eta \\ &\leq \eta^T\left[(\phi+\gamma)G\otimes\left(PA+A^TP+\Gamma\right)-2\sqrt{\rho_0 r_3 G}(\phi+\gamma)\otimes\Gamma\right]\eta.\end{aligned}$$

By choosing a large r_3 such that $r_3 \geq \frac{9g_{\max}}{4\rho_0}$, we have

$$\dot{V}_1(t) \leq \eta^T\left[(\phi+\gamma)G\otimes\left(PA+A^TP-2\Gamma\right)\right]\eta.$$

From condition (10.12), it can be derived that $PA+A^TP-2\Gamma < 0$ with $W = P^{-1}$. Therefore, it is concluded that $\dot{V}_1 < 0$. Hence, the problem is solved.

Remark 10.3.1 *If h_1 is a small enough value such that $0 \leq h_1 \leq h^*$, the predictive error system is asymptotically stable. Obviously, $0 \leq h_1 \leq h^*$ can be satisfied by choosing m large enough. However, the complexity of the controller is proportional to m. If there exists an arbitrary long delay, the controller based on the cascade approach may be involved in an arbitrary long sequence of cascades. This will make the implementation much more complex and degrades the closed-loop system performance. Therefore, the value m should be chosen properly to balance the stability requirement and complexity limitation. Some simulation results will be given in next section to verify the observation.*

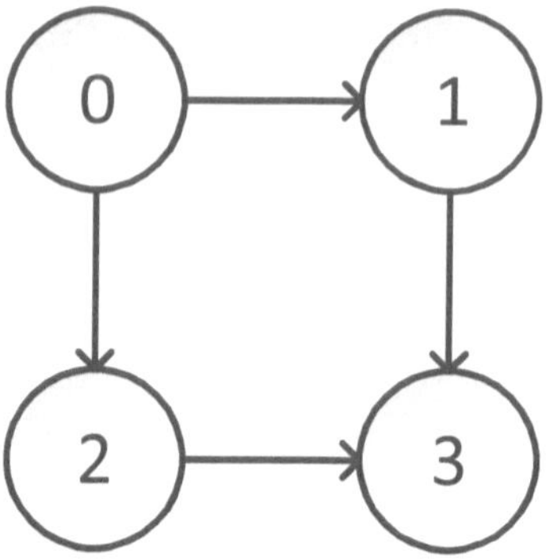

FIGURE 10.1: Communication gragh.

10.4 Numerical Example

Consider a systems with one leader and three followers. The dynamics of the ith agent are described by (10.1), with

$$A = \begin{bmatrix} 0 & 1 \\ -1 & 0 \end{bmatrix}, \quad B = \begin{bmatrix} 0 \\ 1 \end{bmatrix}.$$

Suppose that the initial states of agents are $x_0 = [0;2]$, $x_1 = [-1;2]$, $x_2 = [2;3]$ and $x_3 = [-3;-3]$. The communication gragh is shown in Figure 10.1.

Case 1: System performances with $h = 0.9s$ and $h = 1.2s$

First, we consider the case with $h = 0.9s$. Following the algorithm in Remark 10.2.1, no solution is founded for (10.12) with $h^* = 0.9$. Thus, cascade observer design can be used to solve this problem. Selecting constants $\beta_1 = 4$, $\beta_2 = 1$ and the delay upper bound $h^* = 0.2$, we obtain that

$$P = \begin{bmatrix} 0.7281 & 0.1840 \\ 0.1840 & 0.6881 \end{bmatrix},$$

and the corresponding feedback gains K and Γ are

$$K = \begin{bmatrix} -0.1840 & -0.6881 \end{bmatrix}, \quad \Gamma = \begin{bmatrix} 0.0339 & 0.1266 \\ 0.1266 & 0.4735 \end{bmatrix}.$$

The simulation results are provided in Figure 10.2 with $m = 6$ and $h_1 = 0.15$. Figure 10.2 (a) shows the trajectories of the states. Figure 10.2 (b) shows the observer errors $z(t) - x(t+h)$. Figures 10.2 (c) and 10.2 (d) show the adaptive gains $\gamma + \phi$ and the relative states $z_i(t) - z_i(0)$. From the results above, it can be seen that the consensus is achieved with the proposed cascade observer design method.

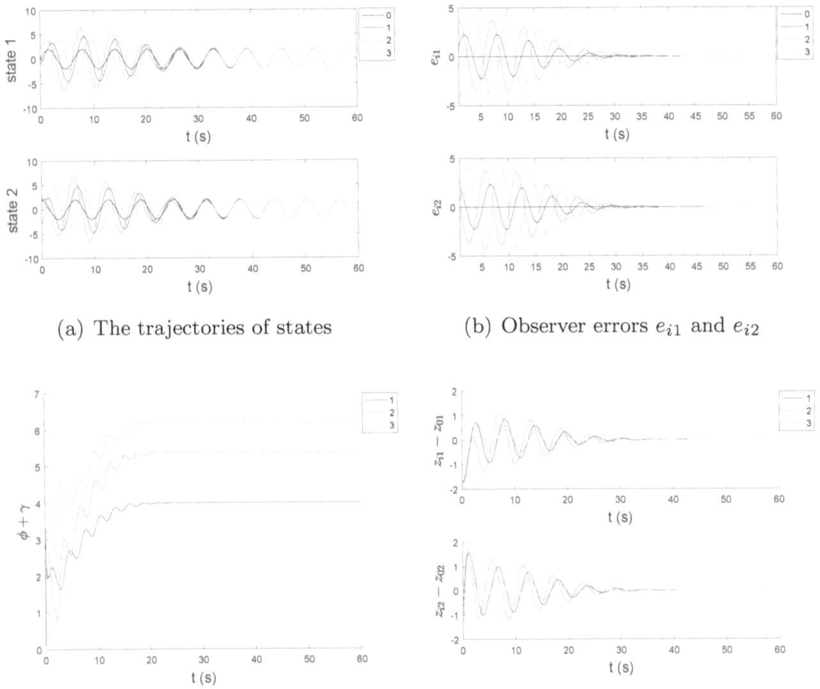

(a) The trajectories of states

(b) Observer errors e_{i1} and e_{i2}

(c) The adaptive gains $\gamma + \phi$

(d) Relative states $z_i - z_0$

FIGURE 10.2: The simulation results with cascade observers ($h = 0.9s$, $m = 6$).

To check the conservatism of (10.12), simulation results with a single predictive observer are also provided in Figure 10.3 with $h_1 = 0.9, m = 1$. It can be seen that even through the single observer-based controller cannot be proved to be stable, it still works and the convergence time is slightly larger than that of the cascade observers.

Then, the systems are tested with $h = 1.2s$. The simulations results are shown in Figure 10.4. It can be seen that the cascade observer still works while the single observer method is no longer effective.

Case 2: System performance with a long sequence of cascade observers

In this case, the effectiveness of long sequence of cascades is tested with a large delay $h = 5s$, In this case, first we choose $m = 10$ and $h_1 = 0.5$. The simulation results are shown in Figure 10.5(a). Then, we choose $m = 25$ and $h_1 = 0.2$. The simulation results are shown in Figure 10.5(b). It proves that the proposed method can be applied to a system with a long delay by using a long sequence of cascades. It should be mentioned that the convergence

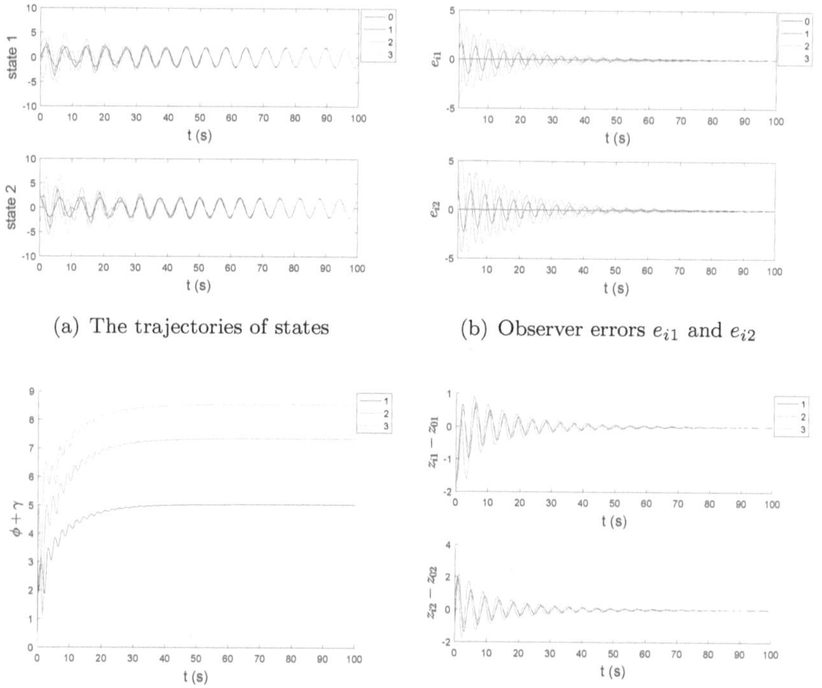

(a) The trajectories of states

(b) Observer errors e_{i1} and e_{i2}

(c) The adaptive gains $\gamma + \phi$

(d) Relative states $z_i - z_0$

FIGURE 10.3: The simulation results with a single observer ($h = 0.9s$, $m = 1$).

performance with $m = 10$ is better than that with $m = 25$. Thus, as mentioned in Remark 10.3.1, m is not as bigger as better.

10.5 Experiment Validation

10.5.1 Experimental Platform

An experimental platform is built to demonstrate the performance of the proposed protocols with real UAV systems. The platform is comprised of four Parrot Bebop quadrotors, the OptiTrack real time tracking systems and a Linux-based computer, as shown in Figure 10.6. In this experiment, we use four quadrotors to achieve a formation flight in the horizontal XY plane under a constant delay. It should be mentioned that for experimental purposes, these quadrotors are treated as 2D fully actuated vehicles with force input. The

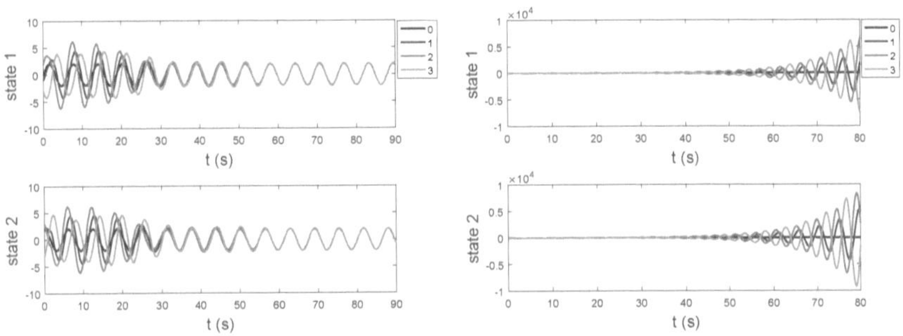

(a) Cascade observer ($h = 1.2$, $m = 8$)

(b) Single observer ($h = 1.2$, $m = 1$)

FIGURE 10.4: The state trajectories under delay $h = 1.2s$ with cascade and single observers.

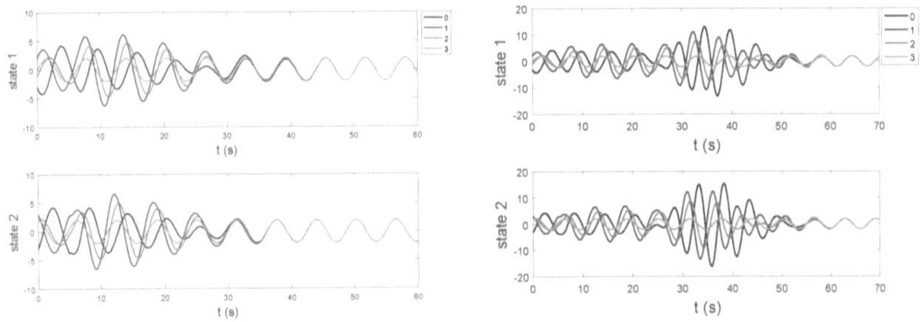

(a) Cascade observers with $h = 5$ and $m = 10$

(b) Cascade observers with $h = 5$ and $m = 25$

FIGURE 10.5: The state trajectories under delay $h = 5$ with different m.

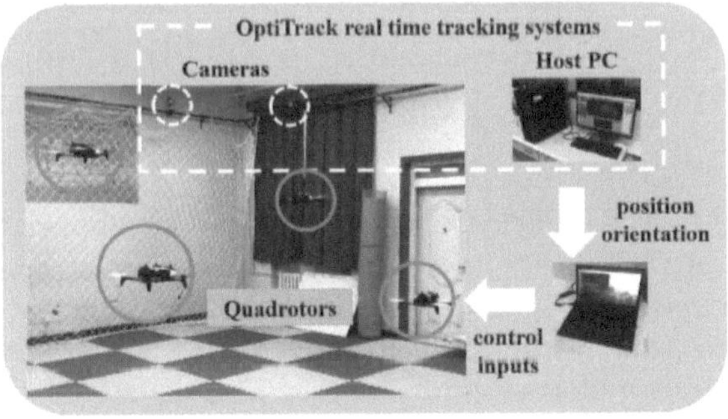

FIGURE 10.6: The multi-quadrotor platform.

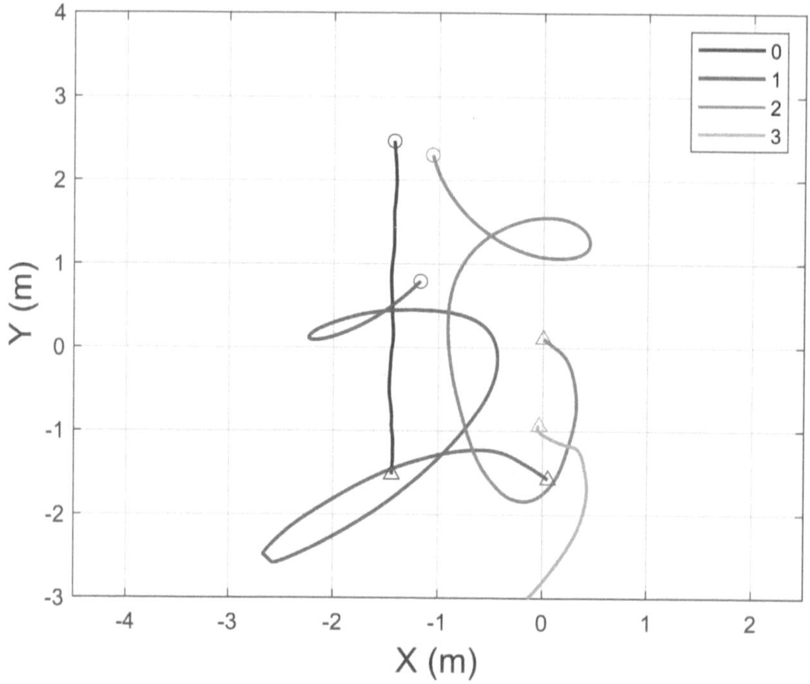

FIGURE 10.7: Position trajectories without predictive observer ($h = 0.6$s).

height and yaw angle of the quadrotor can be controlled separately by using the PID controller [33]. In order to avoid collisions between quadrotors, the heights of the four quadrotors are controlled to be $2.1m$, $1.5m$, $1.8m$ and $1.2m$, respectively. The yaw angles of all quadrotors are controlled to be 0°. By using the linearization method, matrices A and B are given by

$$A = \begin{bmatrix} 0 & 1 \\ 0 & 0 \end{bmatrix}, \; B = \begin{bmatrix} 0 \\ 1 \end{bmatrix}.$$

To characterize the long delay caused by remote communication in outdoor environment, in this experiment, we store the control inputs of quadrotors generated at the current time t, and send the control inputs stored at $(t-h)s$ to each quadrotor. The quadrotor 0 is the leader, which has no control input and maintains uniform motion. Considering the size of the experimental area, the initial states of quadrotors are chosen as $x_0 = [-1.5, 0, -1.5, 0.2]^T$, $x_1 = [0, -0.4, -1.5, 0.2]^T$, $x_2 = [0, 0.2, 0, -0.2]^T$, $x_3 = [0, 0, -0.8, 0]^T$. The application of quadrotors formation flight can be realized by introducing the formation offsets $\phi_i \in \mathbb{R}^4$ to the adaptive protocols (10.21)–(10.23), where ϕ_i denotes the desired separation between quadrotors i and 0. Let

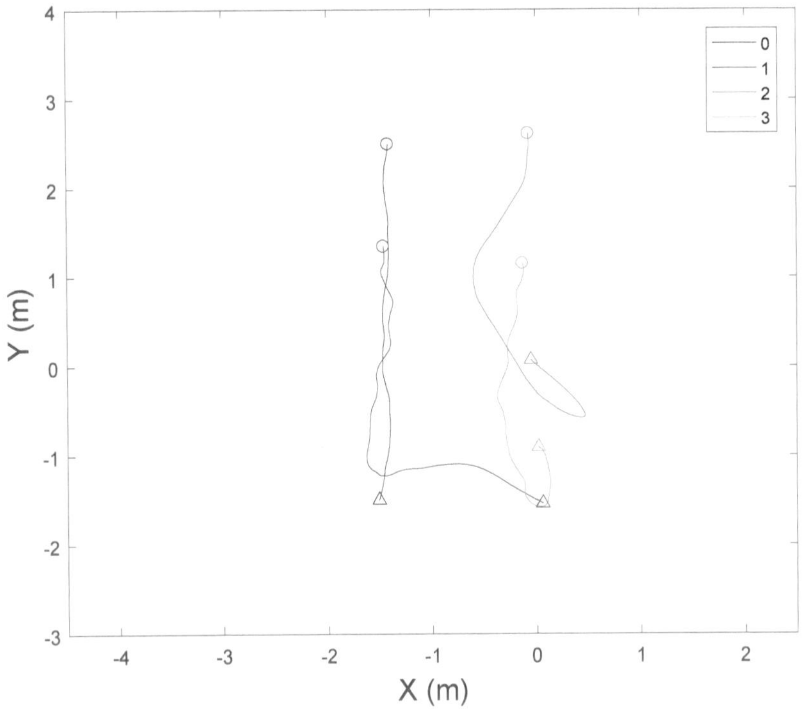

FIGURE 10.8: Position trajectories with predictive observer ($h = 0.6$s).

$\phi_1 = [0, 0, -1.2, 0]^T$, $\phi_2 = [1.2, 0, 0, 0]^T$ and $\phi_3 = [1.2, 0, -1.2, 0]^T$. The desired formation is a rectangle shape.

10.5.2 Experimental Results

To demonstrate the delay effect in formation control, first the experimental results without predictive observer design are shown in Figure 10.7 with $h = 0.6s$, where the initial positions of the four quadrotors are marked by triangles and the final positions are denoted by circles. It can be seen that four quadrotors cannot achieve and maintain a rectangular formation.

Next, the proposed method is applied to compensate the delay effect. According to the analysis, we have a delay upper bound $h^* = 0.15$ with $\beta_1 = 0.1, \beta_2 = 1$, and the feedback gains K and Γ are given as

$$K = [-0.85 - 2.56], \qquad \Gamma = \begin{bmatrix} 0.7296 & 2.1886 \\ 2.1886 & 6.5679 \end{bmatrix}.$$

Figures 10.8 and 10.9 show the position trajectories of the formation flight and the formation geometries of the four quadrotors at time $t \in$

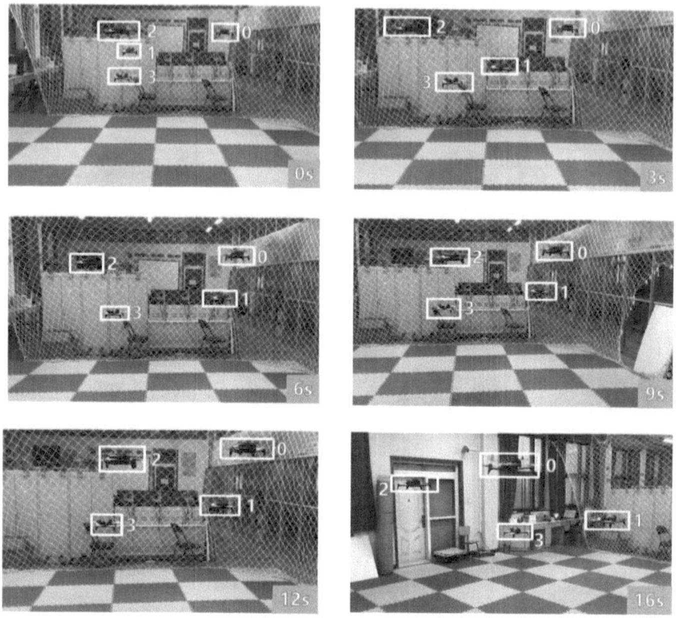

FIGURE 10.9: Formation flight with predictive observer ($h = 0.6$s).

$\{0, 3, 6, 9, 12, 16\}\, s$, respectively. It can be seen that four quadrotors achieve and maintain a rectangular formation.

In the experiment, it is observed that for the small delay case, even though the single observer-based controller cannot be proved to be stable, it still works and the implementation is more simple than that of the cascade observers based controller. However, for the large delay case, the single observer method is no longer effective, while the cascade method can still work. Furthermore, compared with the traditional predictor method, the controllers under this method do not contain any distributed delays and do not need any input information of their neighbours. This benefits the implementation.

10.6 Conclusions

This chapter has investigated the distributed adaptive consensus problem for multi-agent systems with long input delay. Cascade predictive observers are designed for each follower to predict the state in future under a gradual way. The global information of the Laplacian matrix is not needed for each agent.

Consensus analysis is put into the framework of Lyapunov-Krasovskii functionals to guarantee that the consensus errors converge to zero asymptotically. Finally, experimental results have been presented to show the effectiveness of the protocols proposed in this chapter.

10.7 Notes

This chapter extends ideas in [163] and [95] and develops fully distributed consensus protocols for leader-follower multi-agent networks with directed topology. The main materials of this chapter are based on [169]. It is worth mentioning that the consensus protocols proposed in this chapter are designed for linear multi-agent systems. The results in this chapter can be extended to nonlinear multi-agent systems by following the procedures shown in [161,162].

Bibliography

[1] A. Abdessameud and A. Tayebi. Formation control of VTOL unmanned aerial vehicles with communication delays. *Automatica*, 47(11):2383–2394, 2011.

[2] H. Ando, Y. Oasa, I. Suzuki, and M. Yamashita. Distributed memoryless point convergence algorithm for mobile robots with limited visibility. *IEEE Transactions on Robotics and Automation*, 15(5):818–828, 1999.

[3] Z. Artstein. Linear systems with delayed controls: a reduction. *IEEE Transactions on Automatic Control*, 27(4):869–879, 1982.

[4] V. V. Assche, M. Dambrine, J. F. Lafay, and J. P. Richard. Some problems arising in the implementation of distributed-delay control laws. In *38th IEEE Conference on Decision and Control*, pages 4668–4672, Phoenix, Arizona, USA, 1999.

[5] J. Back and J. Kim. A disturbance observer based practical coordinated tracking controller for uncertain heterogeneous multi-agent systems. *International Journal of Robust and Nonlinear Control*, 25(14):2254–2278, 2015.

[6] A. Bahr, J. J. Leonard, and M. F. Fallon. Cooperative localization for autonomous underwater vehicles. *The International Journal of Robotics Research*, 28(6):714–728, 2009.

[7] T. Balch and R. C. Arkin. Behavior-based formation control for multirobot teams. *IEEE Transactions on Robotics and Automation*, 14(6):926–939, 1998.

[8] G. Besançon, D. Georges, and Z. Benayache. Asymptotic state prediction for continuous-time systems with delayed input and application to control. In *European Control Conference (ECC)*, pages 1786–1791, Kos, Greece, 2007.

[9] S. P. Bhat and D. S. Bernstein. Finite-time stability of continuous autonomous systems. *SIAM Journal of Control and Optimization*, 38(3):751–766, 2000.

[10] W. Cao, J. Zhang, and W. Ren. Leader–follower consensus of linear multi-agent systems with unknown external disturbances. *Systems & Control Letters*, 82:64–70, 2015.

[11] Y. Cao and W. Ren. Containment control with multiple stationary or dynamic leaders under a directed interaction graph. In *Proceedings of the 48th IEEE Conference on Decision and Control and the 28th Chinese Control Conference*, 2009.

[12] Y. Cao, D. Stuart, W. Ren, and Z. Meng. Distributed containment control for multiple autonomous vehicles with double-integrator dynamics: algorithms and experiments. *IEEE Transactions on Control Systems Technology*, 19(4):929–938, 2011.

[13] Y. Cao, W. Yu, W. Ren, and G. Chen. An overview of recent progress in the study of distributed multi-agent coordination. *IEEE Transactions on Industrial Informatics*, 9(1):427–438, 2013.

[14] F. Chen, W. Ren, and Z. Lin. Multi-agent coordination with cohesion, dispersion, and containment control. In *Proceedings of the 2011 American Control Conference*, 2010.

[15] H. Chen, J. Wang, C. Wang, J. Shan, and M. Xin. Composite weighted average consensus filtering for space object tracking. *Acta Astronautica*, 168:69–79, 2020.

[16] M. Chen and W. Chen. Disturbance-observer-based robust control for time delay uncertain systems. *International Journal of Control, Automation and Systems*, 8(2):445–453, 2010.

[17] Y. Chen, J. Wang, C. Wang, J. Shan, and M. Xin. Three-dimensional cooperative homing guidance law with field-of-view constraint. *Journal of Guidance, Control, and Dynamics*, 43(2):389–397, 2020.

[18] L. Consolini, F. Morbidi, and D. Prattichizzo. Leader-follower formation control of nonholonomic mobile robots with input constraints. *Automatica*, 44(4):1343–1349, 2008.

[19] P. Cortes, J. Rodriguez, C. Silva, and A. Flores. Delay compensation in model predictive current control of a three-phase inverter. *IEEE Transactions on Industrial Electronics*, 59(2):1323–1325, 2012.

[20] J. Cortés, S. Martinéz, and F. Bullo. Robust rendezvous for mobile autonomous agents via proximity graphs in arbitrary dimensions. *IEEE Transactions on Automatic Control*, 51(8):1289–1298, 2006.

[21] J. P. Desai, J. P. Ostrowski, and V. Kumar. Modeling and control of formations of nonholonomic mobile robots. *IEEE Transactions on Robotics and Automation*, 17(6):905–908, 2001.

[22] D. V. Dimarogonas, M. Egerstedt, and K. J. Kyriakopoulos. A leader-based containment control strategy for multiple unicycles. In *Proceedings of the 45th IEEE Conference on Decision and Control*, 2006.

[23] D. V. Dimarogonas, M. Egerstedt, and K. J. Kyriakopoulos. Further results on the stability of distance-based multi-robot formations. In *Proceedings of the 2009 American Control Conference*, 2009.

[24] Z. Ding. Consensus output regulation of a class of heterogeneous non-linear systems. *IEEE Transactions on Automatic Control*, 58(10):2648–2653, 2013.

[25] Z. Ding. *Nonlinear and Adaptive Control Systems.* IET, London, 2013.

[26] Z. Ding. Consensus control of a class of Lipschitz nonlinear systems. *International Journal of Control*, 87(11):2372–2382, 2014.

[27] Z. Ding. Adaptive consensus output regulation of a class of nonlinear systems with unknown high-frequency gain. *Automatica*, 51:348–355, 2015.

[28] Z. Ding. Consensus disturbance rejection with disturbance observers. *IEEE Transactions on Industrial Electronics*, 62(9):5829–5837, 2015.

[29] Z. Ding and Z. Li. Distributed adaptive consensus control of nonlinear output-feedback systems on directed graphs. *Automatica*, 72(9):46–52, 2016.

[30] Z. Ding and Z. Lin. Truncated state prediction for control of Lipschitz nonlinear systems with input delay. In *Proceeding of IEEE 53rd Annual Conference on Decision and Control (CDC)*, pages 1966–1971. IEEE, 2014.

[31] W. Dong and J. A. Farrell. Cooperative control of multiple non-holonomic mobile agents. *IEEE Transactions on Automatic Control*, 53(6):1434–1448, 2008.

[32] X. Dong, J. Xi, and G. Lu. Formation control for high-order linear time-invariant multiagent systems with time delays. *IEEE Transactions on Control of Network Systems*, 1(3):232–240, 2014.

[33] X. Dong, Y. Zhou, Z. Ren, and Y. Zhong. Time-varying formation tracking for second-order multi-agent systems subjected to switching topologies with application to quadrotor formation flying. *IEEE Transactions on Industrial Electronics*, 64:5014–5024, 2017.

[34] H. Du, S. Li, and P. Shi. Robust consensus algorithm for second-order multi-agent systems with external disturbances. *International Journal of Control*, 85(12):1913–1928, 2012.

[35] H. B. Duan, Q. N. Luo, and Y. X. Yu. Trophallaxis network control approach to formation flight of multiple unmanned aerial vehicles. *Science China Technological Sciences*, 56(4):1066–1074, 2013.

[36] Z. S. Duan, G. R. Chen, and L. Huang. Disconnected synchronized regions of complex dynamical networks. *IEEE Transactions on Automatic Control*, 54(4):845–849, 2009.

[37] K. Engelborghs, M. Dambrine, and D. Roose. Limitations of a class of stabilization methods for delay systems. *IEEE Transactions on Automatic Control*, 46(2):336–339, 2001.

[38] J. A. Fax and R. M. Murray. Graph Laplacians and stabilization of vehicle formations. *IFAC Proceedings Volumes*, 35(1):55–60, 2002.

[39] J. A. Fax and R. M. Murray. Information flow and cooperative control of vehicle formations. *IEEE Transactions on Automatic Control*, 49(9):1465–1476, 2004.

[40] G. Ferrari-Trecate, M. Egerstedt, A. Buffa, and M. Ji. *Laplacian sheep: A hybrid, stop-go policy for leader-based containment control.* Springer, 2006.

[41] D. Fox, J. Ko, K. Konolige, B. Limketkai, D. Schulz, and B. Stewart. Distributed multirobot exploration and mapping. *Proceedings of the IEEE*, 94(7):1325–1339, 2006.

[42] Z. Gao, T. Breikin, and H. Wang. Reliable observer-based control against sensor failures for systems with time delays in both state and input. *IEEE Transactions on Systems, Man, and Cybernetics-Part A: Systems and Humans*, 38(5):1018–1029, 2008.

[43] Z. Gao and S. X. Ding. Actuator fault robust estimation and fault-tolerant control for a class of nonlinear descriptor systems. *Automatica*, 43(5):912–920, 2007.

[44] Z. Gao and H. Wang. Descriptor observer approaches for multivariable systems with measurement noises and application in fault detection and diagnosis. *Systems & Control Letters*, 5(4):304–313, 2006.

[45] K. Gu. An integral inequality in the stability problem of time-delay systems. In *2010 39th IEEE Conference on Decision and Control (CDC)*, volume 2, pages 2805–2810. IEEE, 2010.

[46] K. Gu, J. Chen, and V. L. Kharitonov. *Stability of time-delay systems.* Springer Science & Business Media, 2003.

[47] K. Gu and S.-I. Niculescu. Survey on recent results in the stability and control of time-delay systems. *Journal of Dynamic Systems, Measurement, and Control*, 125(2):158–165, 2003.

[48] Z. Gu, J. H. Park, D. Yue, Z. Wu, and X. Xie. Event-triggered security output feedback control for networked interconnected systems subject to cyber-attacks. *IEEE Transactions on Systems, Man, and Cybernetics: Systems*, 2020.

[49] Z. Gu, P. Shi, D. Yue, S. Yan, and X. Xie. Memory-based continuous event-triggered control for networked TS fuzzy systems against cyber-attacks. *IEEE Transactions on Fuzzy Systems*, 2020.

[50] D. Han and G. Chesi. Robust consensus for uncertain multi-agent systems with discrete-time dynamics. *International Journal of Robust and Nonlinear Control*, 24(13):1858–1872, 2014.

[51] D. Han, G. Chesi, and Y. S. Hung. Robust consensus for a class of uncertain multi-agent dynamical systems. *IEEE Transactions on Industrial Informatics*, 9(1):306–312, 2013.

[52] Y. He, Q.-G. Wang, C. Lin, and M. Wu. Delay-range-dependent stability for systems with time-varying delay. *Automatica*, 43(2):371–376, 2007.

[53] Y. Hong, J. Hu, and L. Gao. Tracking control for multi-agent consensus with an active leader and variable topology. *Automatica*, 42(7):1177–1182, 2006.

[54] R. A. Horn and C. R. Johnson. *Matrix analysis*. Cambridge university press, 2012.

[55] Y. Hu, J. Lam, and J. Liang. Consensus of multi-agent systems with Luenberger observers. *Journal of the Franklin Institute*, 350(9):2769–2790, 2013.

[56] C. Hua, K. Li, and X. Guan. Semi-global/global output consensus for nonlinear multiagent systems with time delays. *Automatica*, 103:480–489, 2019.

[57] A. S. M. Isira, Z. Zuo, and Z. Ding. Leader-follower consensus control of Lipschitz nonlinear systems by output feedback. *International Journal of Systems Science*, 47(16):3772–3781, 2016.

[58] A. Jadbabaie, J. Lin, and A. S. Morse. Coordination of groups of mobile autonomous agents using nearest neighbor rules. *IEEE Transactions on Automatic Control*, 48(6):988–1001, 2003.

[59] M. Ji, G. Ferrari-Trecate, M. Egerstedt, and A. Buffa. Containment control in mobile networks. *IEEE Transactions on Automatic Control*, 53(8):1972–1975, 2008.

[60] Y. Jia. *Robust H_∞ control (in Chinese)*. Beijing: Science Press, Englewood Cliffs, 2007.

[61] H. Kim, H. Shim, and J. Seo. Output consensus of heterogeneous uncertain linear multi-agent systems. *IEEE Transactions on Automatic Control*, 56(1):200–206, 2011.

[62] L. Kocarev and P. Amato. Synchronization in power-law networks. *Chaos: An Interdisciplinary Journal of Nonlinear Science*, 15(2):024101, 2005.

[63] T. Krajník, M. Nitsche, J. Faigl, P. Vaněk, M. Saska, L. Přeučil, T. Duckett, and M. Mejail. A practical multirobot localization system. *Journal of Intelligent & Robotic Systems*, 76(3-4):539–562, 2014.

[64] M. Krstic. *Delay compensation for nonlinear, adaptive, and PDE systems.* Springer, 2009.

[65] M. Krstic. Input delay compensation for forward complete and strict-feedforward nonlinear systems. *IEEE Transactions on Automatic Control*, 55(2):287–303, 2010.

[66] W. Kwon and A. Pearson. Feedback stabilization of linear systems with delayed control. *IEEE Transactions on Automatic Control*, 25(2):266–269, 1980.

[67] D. Lee, Y. Park, and Y. Park. Robust H_∞ sliding mode descriptor observer for fault and output disturbance estimation of uncertain systems. *IEEE Transactions on Automatic Control*, 57(11):2928–2934, 2012.

[68] F. L. Lewis, H. Zhang, K. Hengster-Movric, and A. Das. *Cooperative control of multi-agent systems: optimal and adaptive design approaches.* Springer Science & Business Media, 2013.

[69] J. Li, H. Modares, T. Chai, F. L. Lewis, and L. Xie. Off-policy reinforcement learning for synchronization in multiagent graphical games. *IEEE Transactions on Neural Networks and Learning Systems*, 28(10):2434–2445, 2017.

[70] S. Li, J. Yang, W. Chen, and X. Chen. *Disturbance observer-based control: methods and applications.* CRC press, 2014.

[71] Z. Li and Z. Duan. *Cooperative control of multi-agent systems: A consensus region approach.* CRC Press, 2014.

[72] Z. Li, Z. Duan, and G. Chen. On H_∞ and H_2 performance regions of multi-agent systems. *Automatica*, 47(4):797–803, 2011.

[73] Z. Li, Z. Duan, G. Chen, and L. Huang. Consensus of multi-agent systems and synchronization of complex networks: a unified viewpoint. *IEEE Transactions on Circuits and Systems I: Regular Papers*, 57(1):213–224, 2010.

[74] Z. Li and H. Ishiguro. Consensus of linear multi-agent systems based on full-order observer. *Journal of the Franklin Institute*, 351(2):1151–1160, 2014.

[75] Z. Li, X. Liu, M. Fu, and L. Xie. Global H_∞ consensus of multi-agent systems with Lipschitz non-linear dynamics. *IET Control Theory & Applications*, 6(13):2041–2048, 2012.

[76] Z. Li, W. Ren, X. Liu, and M. Fu. Consensus of multi-agent systems with general linear and Lipschitz nonlinear dynamics using distributed adaptive protocols. *IEEE Transactions on Automatic Control*, 58(7):1786–1791, 2013.

[77] Z. Li, W. Ren, X. Liu, and M. Fu. Distributed containment control for multi-agent systems with general linear dynamics in the presence of multiple leaders. *International Journal of Robust and Nonlinear Control*, 23(5):534–547, 2013.

[78] Z. Li, W. Ren, X. Liu, and L. Xie. Distributed consensus of linear multi-agent systems with adaptive dynamic protocols. *Automatica*, 49(7):1986–1995, 2013.

[79] Z. Li, G. Wen, Z. Duan, and W. Ren. Distributed adaptive consensus control of nonlinear output-feedback systems on directed graphs. *IEEE Transations on Automatic Control*, 60(4):1152–1157, 2015.

[80] J. Liang, Z. Wang, Y. Liu, and X. Liu. State estimation for two-dimensional complex networks with randomly occurring nonlinearities and randomly varying sensor delays. *International Journal of Robust and Nonlinear Control*, 24(1):18–38, 2014.

[81] K.-Y. Liang, S. Van de Hoef, H. Terelius, V. Turri, B. Besselink, J. Mårtensson, and K. H. Johansson. Networked control challenges in collaborative road freight transport. *European Journal of Control*, 30:2–14, 2016.

[82] J. Lin, A. S Morse, and B. D. O. Anderson. The multi-agent rendezvous problem. Part 1: the synchronous case. *SIAM Journal on Control and Optimization*, 46(6):2096–2119, 2007.

[83] J. Lin, A. S Morse, and B. D. O. Anderson. The multi-agent rendezvous problem. Part 2: the asynchronous case. *SIAM Journal on Control and Optimization*, 46(6):2120–2147, 2007.

[84] P. Lin and Y. Jia. Robust H_∞ consensus analysis of a class of second-order multi-agent systems with uncertainty. *IET Control Theory & Applications*, 4(3):487–498, 2010.

[85] P. Lin, Y. Jia, and L. Li. Distributed robust H_∞ consensus control in directed networks of agents with time-delay. *Systems & Control Letters*, 57(8):643–653, 2008.

[86] Z. Lin. *Low gain feedback.* springer London, 1999.

[87] Z. Lin and H. Fang. On asymptotic stabilizability of linear systems with delayed input. *IEEE Transactions on Automatic Control*, 52(6):998–1013, 2007.

[88] Z. Lin, B. Francis, and M. Maggiore. Necessary and sufficient graphical conditions for formation control of unicycles. *IEEE Transactions on Automatic Control*, 50(1):121–127, 2005.

[89] K. Liu, G. Xie, W. Ren, and L. Wang. Consensus for multi-agent systems with inherent nonlinear dynamics under directed topologies. *Systems & Control Letters*, 62:152–162, 2013.

[90] M. Liu and P. Shi. Sensor fault estimation and tolerant control for itô stochastic systems with a descriptor sliding mode approach. *Automatica*, 49(5):1242–1250, 2013.

[91] Y. Liu and Y. Jia. Robust H_∞ consensus control of uncertain multi-agent systems with time delays. *International Journal of Control, Automation and Systems*, 9(6):1086–1094, 2011.

[92] Y. Liu and Y. Jia. H_∞ consensus control for multi-agent systems with linear coupling dynamics and communication delays. *International Journal of Systems Science*, 43(1):50–62, 2012.

[93] J. Lu, J. Cao, and D. W. C. Ho. Adaptive stabilization and synchronization for chaotic Lur'e systems with time-varying delay. *IEEE Transactions on Circuits and Systems I: Regular Papers*, 55(5):1347—1356, 2008.

[94] W. Luand and T. Chen. New approach to synchronization analysis of linearly coupled ordinary differential systems. *Physica D: Nonlinear Phenomena*, 213(2):214–230, 2006.

[95] Y. Lv, Z. Li, Z. Duan, and J. Chen. Distributed adaptive output feedback consensus protocols for linear systems on directed graphs with a leader of bounded input. *Automatica*, 74:308–314, 2016.

[96] A. Manitius and A. Olbrot. Finite spectrum assignment problem for systems with delays. *IEEE Transactions on Automatic Control*, 24(4):541–552, 1979.

[97] S. Martin, A. Girard, A. Fazeli, and A. Jadbabaie. Multiagent flocking under general communication rule. *IEEE Transactions on Control of Network Systems*, 1(2):155–166, 2014.

[98] A. Martinoli, K. Easton, and W. Agassounon. Modeling swarm robotic systems: a case study in collaborative distributed manipulation. *The International Journal of Robotics Research*, 23(4):415–436, 2004.

[99] J. Mei, W. Ren, and G. Ma. Containment control for multiple Euler-Lagrange systems with parametric uncertainties in directed networks. In *Proceedings of the 2011 American Control Conference*, 2011.

[100] Z. Meng, W. Ren, and Z. You. Distributed finite-time attitude containment control for multiple rigid bodies. *Automatica*, 46(2):2092–2099, 2010.

[101] Z. Meng, W.i Ren, Y. Cao, and Z. You. Leaderless and leader-following consensus with communication and input delays under a directed network topology. *IEEE Transactions on Systems, Man, and Cybernetics, Part B (Cybernetics)*, 41(1):75–88, 2010.

[102] M. Mesbahi and F. Y. Hadaegh. Formation flying control of multiple spacecraft via graphs, matrix inequalities, and switching. *Journal of Guidance, Control, and Dynamics*, 24(2):369–377, 2001.

[103] R. R. Murphy, S. Tadokoro, D. Nardi, A. Jacoff, P. Fiorini, H. Choset, and A. M. Erkmen. Search and rescue robotics. In *Springer Handbook of Robotics*, pages 1151–1173. Springer, 2008.

[104] M. Najafi, S. Hosseinnia, F. Sheikholeslam, and M. Karimadini. Closed-loop control of dead time systems via sequential sub-predictors. *International Journal of Control*, 86(4):599–609, 2013.

[105] W. Ni and D. Z. Cheng. Leader-following consensus of multi-agent systems under fixed and switching topologies. *Systems & Control Letters*, 59:209–217, 2010.

[106] H. Nunna and S. Doolla. Multiagent-based distributed-energy-resource management for intelligent microgrids. *IEEE Transactions on Industrial Electronics*, 60(4):1678–1687, 2013.

[107] K. K. Oh, M. C. Park, and H. S. Ahn. A survey of multi-agent formation control. *Automatica*, 53:424–440, 2015.

[108] A. Okubo. Dynamical aspects of animal grouping: swarms, schools, flocks, and herds. *Advances in biophysics*, 22:1–94, 1986.

[109] R. Olfati-Saber. Flocking for multi-agent dynamic systems: Algorithms and theory. *IEEE Transactions on Automatic Control*, 51:401–420, 2006.

[110] R. Olfati-Saber and R. Murray. Consensus problems in networks of agents with switching topology and time-delays. *IEEE Transactions on Automatic Control*, 49(9):1520–1533, 2004.

[111] R. Olfati-Saber and R. M. Murry. Flocking with obstacle avoidance: cooperation with limited communication in mobile networks. In *Proceedings of the 42nd IEEE Conference on Decision and Control*, 2003.

[112] D. A. Paley, F. Zhang, and N. E. Leonard. Cooperative control for ocean sampling: the glider coordinated control system. *IEEE Transactions on Control Systems Technology*, 16(4):735–744, 2008.

[113] Z. J. Palmor. Time-delay compensation-Smith predictor and its modifications. *The control handbook*, 1:224–229, 1996.

[114] M. N. A. Parlakçı. Improved robust stability criteria and design of robust stabilizing controller for uncertain linear time-delay systems. *International Journal of Robust and Nonlinear Control*, 16(13):599–636, 2006.

[115] L. M. Pecora and T. L. Carroll. Synchronization in chaotic systems. *Physical Review Letters*, 64(8):821–824, 1990.

[116] L. M. Pecora and T. L. Carroll. Master stability functions for synchronized coupled systems. *Physical Review Letters*, 80(10):2109–2112, 1998.

[117] L. M. Pecora and T. L. Carroll. Chaos synchronization in complex networks. *IEEE Transactions on Circuits and Systems I: Regular Papers*, 55(5):1135–1146, 2008.

[118] H. A. Poonawala, A. C. Satici, H. Eckert, and M. W. Spong. Collision-free formation control with decentralized connectivity preservation for nonholonomic-wheeled mobile robots. *IEEE Transactions on Control of Network Systems*, 2(2):122–130, 2015.

[119] M. Porfiri, D. J. Stilwell, and E. M. Bollt. Synchronization in random weighted directed networks. *IEEE Transactions on Circuits and Systems I: Regular Papers*, 55(10):3170–3177, 2008.

[120] A. Poznyak, A. Polyakov, and V. Azhmyakov. *Attractive Ellipsoids in Robust Control*. Birkhäuser: Springer, 2014.

[121] W. Qiao and R. Sipahi. Consensus control under communication delay in a three-robot system: Design and experiments. *IEEE Transactions on Control Systems Technology*, 24(2):687–694, 2016.

[122] Z. Qu. *Cooperative control of dynamical systems: applications to autonomous vehicles*. Springer Science & Business Media, 2009.

[123] W. Ren. Information consensus in multi-vehicle cooperative control. *IEEE Control Systems Magazine*, 27(2):71–82, 2007.

[124] W. Ren and R. Beard. Decentralized scheme for spacecraft formation flying via the virtual structure approach. *Journal of Guidance, Control, and Dynamics*, 27(1):73–82, 2004.

[125] W. Ren and R. W. Beard. Consensus seeking in multiagent systems under dynamically changing interaction topologies. *IEEE Transactions on Automatic Control*, 50(5):655–661, 2005.

[126] W. Ren and R. W. Beard. Consensus algorithms for double-integrator dynamics. *Distributed Consensus in Multi-vehicle Cooperative Control: Theory and Applications*, pages 77–104, 2008.

[127] W. Ren and R. W. Beard. *Distributed consensus in multi-vehicle cooperative control.* Springer, 2008.

[128] W. Ren and N. Sorensen. Distributed coordination architecture for multi-robot formation control. *Robotics and Autonomous Systems*, 56(4):324–333, 2008.

[129] C. W. Reynolds. Flocks, herds and schools: a distributed behavioral model. In *ACM SIGGRAPH computer graphics*, volume 21, pages 25–34. ACM, 1987.

[130] J.-P. Richard. Time-delay systems: an overview of some recent advances and open problems. *Automatica*, 39(10):1667–1694, 2003.

[131] C. Sabol, R. Burns, and C. A. McLaughlin. Satellite formation flying design and evolution. *Journal of Spacecraft and Rockets*, 38(2):270–278, 2001.

[132] I. Saboori and K. Khorasani. H_∞ consensus achievement of multi-agent systems with directed and switching topology networks. *IEEE Transactions on Automatic Control*, 59(11):3104–3109, 2014.

[133] J. A. Sauter, R. Matthews, H. Van D Parunak, and S. A. Brueckner. Performance of digital pheromones for swarming vehicle control. In *Proceedings of the fourth international joint conference on Autonomous agents and multiagent systems*, pages 903–910. ACM, 2005.

[134] L. Schenato and G. Gamba. A distributed consensus protocol for clock synchronization in wireless sensor network. In *proceeding of 46th IEEE Conference on Decision and Control*, pages 2289–2294. IEEE, 2007.

[135] R. Simmons, D. Apfelbaum, W. Burgard, D. Fox, M. Moors, S. Thrun, and H. Younes. Coordination for multi-robot exploration and mapping. In *AAAI/IAAI*, pages 852–858, 2000.

[136] O. J. M. Smith. A controller to overcome dead time. *ISA Journal*, 6(2):28–33, 1959.

[137] W. Song, J. Wang, C. Wang, and J. Shan. A variance-constrained approach to event-triggered distributed extended Kalman filtering with multiple fading measurements. *International Journal of Robust and Nonlinear Control*, 29(5):1558–1576, 2019.

[138] D. M. Stipanović, G. Inalhan, R. Teo, and C. J. Tomlin. Decentralized overlapping control of a formation of unmanned aerial vehicles. *Automatica*, 40(8):1285–1296, 2004.

[139] H. Su, X. Wang, and Z. Lin. Flocking of multi-agents with a virtual leader. *IEEE Transactions on Automatic Control*, 54(2):293–307, 2009.

[140] Y. Su and J. Huang. Cooperative adaptive output regulation for a class of nonlinear uncertain multi-agent systems with unknown leader. *Systems & Control Letters*, 62(6):461–467, 2013.

[141] C. Sun, H. Duan, and Y. Shi. Optimal satellite formation reconfiguration based on closed-loop brain storm optimization. *IEEE Computational Intelligence Magazine*, 8(4):39–51, 2013.

[142] J. Sun, Z. Geng, and Y. Lv. Adaptive output feedback consensus tracking for heterogeneous multi-agent systems with unknown dynamics under directed graphs. *Systems & Control Letters*, 87:16–22, 2016.

[143] J. Sun, Z. Geng, Y. Lv, Z. Li, and Z. Ding. Distributed adaptive consensus disturbance rejection for multi-agent systems on directed graphs. *IEEE Transations on Control of Network Systems*, 5(1):629–639, 2016.

[144] X. Sun, Z. Gu, F. Yang, and S. Yan. Memory-event-trigger-based secure control of cloud-aided active suspension systems against deception attacks. *Information Sciences*, 543:1–17, 2021.

[145] Y. Sun and L. Wang. Consensus of multi-agent systems in directed networks with nonuniform time-varying delays. *IEEE Transactions on Automatic Control*, 54(7):1607–1613, 2009.

[146] H. G. Tanner, A. Jadbabaie, and G. J. Pappas. Stable flocking of mobile agents, Part I: Fixed topology. In *Proceedings of the 42nd IEEE Conference on Decision and Control*, 2003.

[147] H. G. Tanner, A. Jadbabaie, and G. J. Pappas. Stable flocking of mobile agents Part II: Dynamic topology. In *Proceedings of the 42nd IEEE Conference on Decision and Control*, 2003.

[148] H. G. Tanner, A. Jadbabaie, and G. J. Pappas. Flocking in fixed and switching networks. *IEEE Transactions on Automatic Control*, 52(5):863–868, 2007.

[149] Y. Tian and C. Liu. Robust consensus of multi-agent systems with diverse input delays and asymmetric interconnection perturbations. *Automatica*, 45(5):1347–1353, 2009.

[150] Y. Tian and Y. Zhang. High-order consensus of heterogeneous multi-agent systems with unknown communication delays. *Automatica*, 48(6):1205–1212, 2012.

[151] A. Tiwari, J. Fung, and J. M. Carson. A framework for Lyapunov certificates for multi-vehicle rendezvous problem. In *Proceedings of the 2004 American Control Conference*, 2004.

[152] T. Vicsek, A. Czirók, E. Ben-Jacob, I. Cohen, and O. Shochet. Novel type of phase transition in a system of self-driven particles. *Physical Review Letters*, 75(6):1226–1229, 1995.

[153] C. Wang. Consensus control for multi-agent sytems with input delay. *PhD Thesis*, 2015.

[154] C. Wang, X. Ding, J. Wang, and J. Shan. A robust three-dimensional cooperative guidance law against maneuvering target. *Journal of the Franklin Institute*, 357(10):5735–5752, 2020.

[155] C. Wang and Z. Ding. H_∞ consensus control of multi-agent systems with input delay and directed topology. *IET Control Theory & Applications*, 10(6):617–624, 2016.

[156] C. Wang, W. Dong, J. Wang, and Z. Ding. Predictive descriptor observer design for a class of LTI systems with applications to quadrotor trajectory tracking. *IEEE Transactions on Industrial Electronics, online published.*

[157] C. Wang, H. Tnunay, Z. Zuo, B. Lennox, and Z. Ding. Fixed-time formation control of multi-robot systems: Design and experiments. *IEEE Transactions on Industrial Electronics*, 66(8):6292–6301, 2019.

[158] C. Wang, J. Yang, Z. Zuo, and Z. Ding. Output feedback disturbance rejection for a class of linear systems with input delay via dobc approach. In *2016 35th Chinese Control Conference (CCC)*, pages 136–141. IEEE, 2016.

[159] C. Wang, Z. Zuo, and Z. Ding. A control scheme for LTI systems with Lipschitz nonlinearity and unknown time-varying input delay. *IET Control Theory & Applications*, 11(17):3191–3195, 2017.

[160] C. Wang, Z. Zuo, Q. Gong, and Z. Ding. Formation control with disturbance rejection for a class of Lipschitz nonlinear systems. *Science China Information Sciences*, 60(7):070202, 2017.

[161] C. Wang, Z. Zuo, Z. Lin, and Z. Ding. Consensus control of a class of Lipschitz nonlinear systems with input delay. *IEEE Transactions on Circuits and Systems I: Regular Papers*, 62(11):2730–2738, 2015.

[162] C. Wang, Z. Zuo, Z. Lin, and Z. Ding. A truncated prediction approach to consensus control of Lipschitz nonlinear multi-agent systems with input delay. *IEEE Transactions on Control of Network Systems*, 4(4):716–724, 2017.

[163] C. Wang, Z. Zuo, Z. Qi, and Z. Ding. Predictor-based extended-state-observer design for consensus of MASs with delays and disturbances. *IEEE Transactions on Cybernetics*, 49:1259–1269, 2019.

[164] C. Wang, Z. Zuo, J. Sun, J. Yang, and Z. Ding. Consensus disturbance rejection for Lipschitz nonlinear multi-agent systems with input delay: A DOBC approach. *Journal of the Franklin Institute*, 354(1):298–315, 2017.

[165] C. Wang, Z. Zuo, Y. Wu, and Z. Ding. Fixed-time nonlinear consensus algorithms for multi-agent systems with input delay. In *2017 13th IEEE International Conference on Control & Automation (ICCA)*, pages 52–57. IEEE, 2017.

[166] D. Wang, N. Zhang, J. Wang, and W. Wang. A PD-like protocol with a time delay to average consensus control for multi-agent systems under an arbitrarily fast switching topology. *IEEE Transactions on Cybernetics*, 47(4):898–907, 2016.

[167] J. Wang, Z. Duan, Z. Li, and G. Wen. Distributed H_∞ and H_2 consensus control in directed networks. *IET Control Theory & Applications*, 8(3):193–201, 2014.

[168] J. Wang, C. Wang, M. Xin, Z. Ding, and J. Shan. *Cooperative Control of Multi-Agent Systems: An Optimal and Robust Perspective.* Academic Press, 2020.

[169] J. Wang, Z. Zhou, C. Wang, and Z. Ding. Cascade structure predictive observer design for consensus control with applications to UAVs formation flying. *Automatica*, 121:109200, 2020.

[170] J. Wang, Z. Zhou, C. Wang, and J. Shan. Multiple quadrotors formation flying control design and experimental verification. *Unmanned Systems*, 7(1):47–54, 2019.

[171] X. Wang and S. Li. Nonlinear consensus algorithms for second-order multi-agent systems with mismatched disturbances. In *Proceeding of 2015 American Control Conference (ACC)*, pages 1740–1745, July 2015.

[172] X. Wang, G. Zhong, K. Tang, K. Man, and Z. Liu. Generating chaos in Chua's circuit via time-delay feedback. *IEEE Transactions on Circuits and Systems I: Regular Papers*, 48(9):1151–1156, 2001.

[173] X. F. Wang and G. R. Chen. Synchronization in scale-free dynamical networks: Robustness and fragility. *IEEE Transactions on Circuits and Systems I: Fundamental Theory and Applications*, 49(1):54–62, 2002.

[174] Y. Wei and Z. Lin. Delay independent truncated predictor feedback for stabilization of linear systems with multiple time-varying input delays. In *2017 American Control Conference (ACC)*, Seattle, WA, USA, pages 5732–5737. IEEE, 2017.

[175] G. Wen, Z. Duan, G. Chen, and W. Yu. Consensus tracking of multi-agent systems with Lipschitz-type node dynamics and switching topologies. *IEEE Transactions on Circuits and Systems I: Regular Papers*, 61(2):499–511, 2014.

[176] G. Wen, Z. Duan, W. Yu, and G. Chen. Consensus in multi-agent systems with communication constraints. *International Journal of Robust and Nonlinear Control*, 22(2):170–182, 2012.

[177] G. Wen, G. Hu, W. Yu, J. Cao, and G. Chen. Consensus tracking for higher-order multi-agent systems with switching directed topologies and occasionally missing control inputs. *Systems & Control Letters*, 62(12):1151–1158, 2013.

[178] G. Wen, G. Hu, W. Yu, and G. Chen. Distributed H_∞ consensus of higher order multiagent systems with switching topologies. *IEEE Transactions on Circuits and Systems II: Express Briefs*, 61(5):359–363, 2014.

[179] G. Wen, W. Yu, M.Z.Q. Chen, X. Yu, and G. Chen. H_∞ pinning synchronization of directed networks with aperiodic sampled-data communications. *IEEE Transactions on Circuits and Systems I: Regular Papers*, 61(11):3245–3255, 2014.

[180] W. Wu, W. Zhou, and T. Chen. Cluster synchronization of linearly coupled complex networks under pinning control. *IEEE Transactions on Circuits and Systems I: Regular Papers*, 56(4):829–839, 2009.

[181] Y. Wu, H. Su, P. Shi, Z. Shu, and Z. G. Wu. Consensus of multiagent systems using aperiodic sampled-data control. *IEEE Transactions on Cybernetics*, 46(9):2132–2143, 2016.

[182] S. Xu and J. Lam. Improved delay-dependent stability criteria for time-delay. *IEEE Transactions on Automatic Control*, 50(3):384–387, 2005.

[183] A. Yamashita, M. Fukuchi, J. Ota, T. Arai, and H. Asama. Motion planning for cooperative transportation of a large object by multiple

mobile robots in a 3D environment. In *Proceedings of IEEE International Conference on ICRA*, pages 3144–3151. IEEE, 2000.

[184] H. Yang, Z. Zhang, and S. Zhang. Consensus of second-order multi-agent systems with exogenous disturbances. *International Journal of Robust and Nonlinear Control*, 21(9):945–956, 2011.

[185] W. Yang, X. Wang, and H. Shi. Fast consensus seeking in multi-agent systems with time delay. *Systems & Control Letters*, 62(3):269–276, 2013.

[186] X. Yang, K. Watanabe, K. Izumi, and K. Kiguchi. A decentralized control system for cooperative transportation by multiple non-holonomic mobile robots. *International Journal of Control*, 77(10):949–963, 2004.

[187] S. Yin, H. Gao, J. Qiu, and O. Kaynak. Descriptor reduced-order sliding mode observers design for switched systems with sensor and actuator faults. *Automatica*, 76(5):282–292, 2017.

[188] S. Y. Yoon, P. Anantachaisilp, and Z. Lin. An LMI approach to the control of exponentially unstable systems with input time delay. In *Proceeding of IEEE 52nd Annual Conference on Decision and Control (CDC)*, pages 312–317. IEEE, 2013.

[189] S. Y. Yoon and Z. Lin. Predictor based control of linear systems with state, input and output delays. *Automatica*, 53:385–391, 2015.

[190] W. Yu, G. Chen, and M. Cao. Consensus in directed networks of agents with nonlinear dynamics. *IEEE Transactions on Automatic Control*, 56(6):1436–1441, 2011.

[191] W. Yu, G. Chen, M. Cao, and J. Kurths. Second-order consensus for multiagent systems with directed topologies and nonlinear dynamics. *IEEE Transactions on Systems, Man, and Cybernetics, Part B*, 40(3):881–891, 2010.

[192] H. Zhang, D. Yue, C. Dou, W. Zhao, and X. Xie. Data-driven distributed optimal consensus control for unknown multiagent systems with input-delay. *IEEE Transactions on Cybernetics*, 49(6):2095–2105, 2018.

[193] W.-A. Zhang and L. Yu. Stability analysis for discrete-time switched time-delay systems. *Automatica*, 45(10):2265–2271, 2009.

[194] X. Zhang, H. Duan, and Y. Yu. Receding horizon control for multi-uavs close formation control based on differential evolution. *Science China Information Science*, 53(4):223–235, 2010.

[195] Z. Zhao and Z. Lin. Global leader-following consensus of a group of general linear systems using bounded controls. *Automatica*, 68:294–304, 2016.

[196] Y. Zheng, Y. Zhu, and L. Wang. Consensus of heterogeneous multi-agent systems. *IET Control Theory & Applications*, 5(16):1881–1888, 2011.

[197] Q. C. Zhong. *Robust control of time-delay systems.* Springer Science & Business Media, 2006.

[198] B. Zhou and Z. Lin. Consensus of high-order multi-agent systems with large input and communication delays. *Automatica*, 50(2):452–464, 2014.

[199] B. Zhou, Z. Lin, and G. R. Duan. Stabilization of linear systems with input delay and saturation - a parametric Lyapunov equation approach. *International Journal of Robust and Nonlinear Control*, 20(4):1502–1519, 2010.

[200] B. Zhou, Z. Lin, and G. R. Duan. Truncated predictor feedback for linear systems with long time-varying input delays. *Automatica*, 48(10):2387–2399, 2012.

[201] B. Zhou, Q. Liu, and F. Mazenc. Stabilization of linear systems with both input and state delays by observer–predictors. *Automatica*, 83:368–377, 2017.

[202] K. Zhou and J. C. Doyle. *Essentials of robust control*, volume 104. NJ: Prentice hall, 1998.

[203] Z. Zuo. Nonsingular fixed-time consensus tracking for second-order multi-agent networks. *Automatica*, 54(11):305–309, 2015.

[204] Z. Zuo, M. Defoort, B. Tian, and Z. Ding. Distributed consensus observer for multi-agent systems with high-order integrator dynamics. *IEEE Transactions on Automatic Control*, 65(4):1771–1778, 2020.

[205] Z. Zuo, Q.-L. Han, B. Ning, X. Ge, and X.-M. Zhang. An overview of recent advances in fixed-time cooperative control of multi-agent systems. *IEEE Transactions on Industrial Informatics*, 14(6):2322–2334, 2018.

[206] Z. Zuo, Z. Lin, and Z. Ding. Prediction output feedback control of a class of Lipschitz nonlinear systems with input delay. *IEEE Transactions on Circuits and Systems II: Express Briefs*, 63(8):788–792, 2016.

[207] Z. Zuo, Z. Lin, and Z. Ding. Truncated predictor control of Lipschitz nonlinear systems with time-varying input delay. *IEEE Transactions on Automatic Control*, 62(10):5324–5330, 2017.

[208] Z. Zuo and L. Tie. A new class of finite-time nonlinear consensus protocols for multi-agent systems. *International Journal of Control*, 87(2):363–370, 2014.

[209] Z. Zuo, C. Wang, and Z. Ding. A reduction method to consensus control of uncertain multi-agent systems with input delay. In *Proceeding of 34th Chinese Control Conference (CCC)*, pages 7315–7321. IEEE, 2015.

[210] Z. Zuo, C. Wang, and Z. Ding. Robust consensus control of uncertain multi-gent systems with input delay: a model reduction method. *International Journal of Robust and Nonlinear Control*, 27(11):1874–1894, 2017.

[211] Z. Zuo, C. Wang, W. Yang, and Z. Ding. Robust $\mathcal{L}_2$ disturbance attenuation for a class of uncertain Lipschitz nonlinear systems with input delay. *International Journal of Control*, 92(5):1015–1021, 2019.

[212] Z. Zuo, W. Yang, L. Tie, and D. Meng. Fixed-time consensus for multi-agent systems under directed and switching interaction topology. In *Proceeding of 2014 American Control Conference (ACC)*, pages 5133–5138, 2014.